Great Lakes Nature

AN OUTDOOR YEAR

Mary Blocksma

Illustrations by ROBIN WILT

THE UNIVERSITY OF MICHIGAN PRESS

Ann Arbor

Library of Congress Cataloging-in-Publication Data

Blocksma, Mary.
  [Naming nature]
  Great Lakes nature : an outdoor year / Mary Blocksma ; illustrations by Robin Wilt.
     p.   cm.
  Originally published: Naming nature. New York, N.Y. : Penguin Books, 1992.
  Includes bibliographical references (p.   ).
  ISBN 0-472-08929-3 (pbk. : alk. paper)
     1. Zoology—Nomenclature (Popular)—Study and teaching. 2. Plant names,
  Popular—Study and teaching. 3. Natural history—Great Lakes Region. I. Title.

  QL355.B52   2003
  508.77—dc21

                                                                2003042669

Cover photo by Roger Eriksson, copyright © Roger Eriksson

To Ellen Wilt,
who plied me with feasts for body,
book and soul

A Note About the Title

This book was first published by Penguin USA as NAMING NATURE.
I have changed the title of this new edition to GREAT LAKES NATURE,
a more accurate and helpful description of the outdoor delights that follow.

—M.B.

Acknowledgments

This book could not have been written without the unfailing support of many. First in line for my gratitude is Mother Nature, who supplied me with endless material, entertainment and special performances, from the bloom of a thousand petals to eagles flung past my windows.

Very special thanks are in order to my parents, Ralph and Ruth Blocksma, for supporting every wild idea I have ever come up with; to the two Marys (besides myself, of course) who appear frequently in this book, Mary Heuvelhorst in Douglas and Mary Stewart Scholl on Beaver Island, for their inspiring familiarity with and love of nature; to Al Vileisis, for his companionship, considerable knowledge and his canoe; and to Ellen Wilt, the illustrator's mother and my dear friend, to whom I have dedicated this book.

My thanks are also due to the many other cheerful participants, including Mary Brodbeck, Lyn Coffin, Judy Hallisy, Doug Hagley, Joan Heuvelhorst, Russell Hibbard, Howard and Sally Hunt, Marcia Perry, Marchiene (Marti) Rienstra and Bette Williams, who appear in the Douglas-Saugatuck portions of the book; and on Beaver Island, Jon Barrett, Carol Hart, Bill Freese, Roy Elsworth, Glen Felix, Eric Heline, Don Meister (also known as Wassakwaam), Judi Meister and Cindy Ricksgers. My heartfelt thanks also to all persons I encountered in the writing of this book whose names I never knew or—please forgive me—who may have slipped my mention.

I owe much to the experts, all but the last two from Michigan, whom I consulted on many occasions and who donated much of their valuable time: Jim Bruton, astronomer at the Jesse Besser Museum in Alpena; Jim Kane, chairperson, Life Sciences Department, Muskegon Community College, Muskegon; Keewaydinoquay, medicine woman and botanist from Leland; Deborah Torres, director, Saugatuck-Douglas District Library; and Earl Wolf, naturalist at the Gillette Nature Center, Hoffmaster State Park, Muskegon. I also thank my cousin Richard Greaves from Great Britain and Taylor Schoettle of the University of Georgia Marine Extension Service. None of the above are in any way responsible for any errors I may have made in this book.

And, last but best, I cannot exaggerate my gratitude to my agent, Gina Maccoby, for her tireless, cheerful support; to Robin Wilt, who put so much heart and skill into the art and cared whether or not I liked it; and to Mindy Werner, my patient, thorough and exceptionally kind Penguin USA editor, who, together with designer Kate Nichols, transformed my manuscript into this book.

Contents

Introduction

> "There are moments in our lives, there are moments in a day, when we see beyond the usual. Such are the moments of our greatest happiness. Such are the moments of our greatest wisdom."
>
> —Robert Henri, *The Art of Spirit*

One day I was gazing out my window at a stretch of trees when I was suddenly struck with the realization that I couldn't name any of them. It was as if a pair of window shades had, without warning, rolled themselves up with a snap. That moment, that clap of light, changed my life. I became conscious, for the first time, of how little I knew of my natural environment, and I was appalled. I loved nature. The daughter of medical missionaries, I had spent much of my childhood chasing parrots on the plains of Pakistan and butterflies and monkeys in the thickly wooded foothills of the Himalayas. Still, like most Americans, I had somehow become an adult who could not claim even an elementary knowledge of my natural neighborhood. How, I wondered, will we save the world if we can't even name our nearest trees?

I decided to make their acquaintance. I would learn their names as well as those of my other wild neighbors, but where to begin? Nature's enormity, complexity and variety were overwhelming, until it occurred to me that if I just could name the most common plants, birds and animals, I'd probably recognize half of what I saw. On this framework, I could hang, like ornaments, the more unusual species I found.

Having carved nature in half, I already felt more hopeful. I decided to make it as easy on myself as possible. I approached my naming nature project whimsically, suiting myself, naming anything I pleased—no big program, no particular organization, just a name at a time, a few times a week, for a year. (When a dead language nearly frightened me out of the living world, I left Latin names to the guidebooks.) I would have fun with this, and if it stopped being fun, I'd quit. In a surprisingly short time— only a few weeks—I'd been drawn into the game, my enthusiasm growing with my "natural" vocabulary. By April, encouraged by friends who wanted to be able to name nature, too, I realized that what had begun as a New Year's resolution might very well make a book. I'd kept a journal, and drawing on my expertise as a librarian, I could add researched material that would make the naming even easier for my readers than it had been for me.

Sharing my experiences in naming nature sounded more and more like a good idea. Much has been made of cultural literacy, but it seemed to me that natural literacy was at least as critical, having been lost so long ago that even educated persons like my friends and myself were unaware of how little we knew. Our parents hadn't taught us, and their parents hadn't taught them. I expected that there were many others who, like me, would not miss what they did not see, and would not see what they could not name.

Familiarity breeds intimacy, and intimacy, concern. It is hard for me to care about anything—person, place or thing—that I cannot name. Naming has been for me a powerful way to take into myself something or someone important to me. It makes the unknown friendly, approachable, knowable, sometimes even lovable. Before I began this year, I experienced Nature afar as "landscape," up close as "ambiance" and in my conscience as "The Environment." Entering a woods was like moving into a neighborhood where I knew almost no one. Now I have befriended so many inhabitants that if any of them moved away, I would know and feel bereft.

Naming is not the only way to appreciate the natural world, but it can be a gentle, playful way to say hello. It's an entrance available to anyone, anywhere. All that is required is a few walks a week, a few basic nature guides, some note-taking or sketches and possibly a pair of binoculars. I found nature surprisingly eager to entertain anyone, even a rank amateur, who shows up with some regularity. It is not boring out there. Each person's discoveries and adventures will be, like mine, unique, depending on his or her perception and corner of the world. (Naturally, anyone wishing to explore private property will, as I did, obtain permission first.)

As for the book, I confined myself to the United States and Canada. Many of the plants, animals and birds named here are common all over

North America. Because I live in Michigan, more are found in the eastern half, and all in the northeast quarter. I've included a rough range estimate for most species. And I've tried to make helpful comparisons and observations often not found in nature guides. But as a retired biology professor said to me once, "I always tell my students that everything I tell them is probably only 90 percent true." Nature, like language, has a way of insisting on exceptions to the rules.

With that, welcome to your world. You can explore my part of it from your chair, or befriend your own as well. All I ask is that the next time you go out, notice the trees. Do you know, specifically, what they are? Have you met yet, been properly introduced? Allow me.

January

LAKE SUPERIOR

CANADA

MICHIGAN

Mackinac
Bridge

Beaver
Island

Charlevoix

LAKE
HURON

LAKE MICHIGAN

Robin

MICHIGAN

Grand Rapids

Holland
Saugatuck
Douglas

Lansing

South Haven

Detroit

CANADA

Chicago

LAKE
ERIE

A New Year's Resolution

It's hard for me to believe this, but I can't name the trees—not *one* of the trees—that are growing outside my windows. There are bare trees and there are evergreens, both tall and short—tree and shrub?—but that is as much as I know. How, I wonder, could I have lived nearly fifty years and know so little, and why am I only now becoming conscious of my "natural illiteracy"? The answers to those questions have to do with a culture that hasn't cared much about what's what in the environment, to the point where even an educated person and nature lover like me can go a lifetime unaware of the extent of my ignorance.

I have just moved out of California and into an apartment on the second floor of a house designed by a student of Frank Lloyd Wright, built into the side of a sand dune on the eastern shore of Lake Michigan. A bank of north- and west-facing windows wrap halfway around a hexagonal living room, giving me a cinematic view of the middles of trees, evergreen boughs, deserted summer cottages and lake. I stared for hours this afternoon at the woods and snow outside my windows. I couldn't find one thing I knew.

I am resolving, therefore, to become naturally literate. I will try to name several things each week, for the next year, using the days in between to practice what I've already learned. This resolution fits in well with my reasons for having moved here: I am here to heal, to begin a new life after divorce. Becoming closer to the natural world sounds comforting to me. There are other reasons, too. As a creative person, I crave new ways of seeing. I know I don't really "see" what's right in front of my nose, and naming things may help sharpen my vision. There is also my concern for, but relative ignorance of, the environment. How can I ever really care, when I know so little about even my neighborhood? So this is the year I will learn to name the trees, at least the ones around me, and some birds and flowers, too. Maybe even a few ferns and mushrooms.

Lake Effect Snow

All right, I've made a noble commitment. Now what? Where does a person begin naming nature? Well, the longest journey begins with one step. If I average several names a week, what does it matter what they are? A name here, a name there, and I'll soon be on speaking terms with the most frequently appearing parts of my environment. As for where to begin, I consider the old writer's adage: Write What You Know. I've never thought much of this advice, preferring to write about things I don't know, thus educating myself. But feeling vastly intimidated by the entire world of nature out there, I might be wise to start where I know at least something.

For two weeks, since I've moved here, I've shoveled the driveway every day, enjoying the quiet company of the snow-filled woods. Snow is something that I *can* name: This isn't, I know, just any snow. This is lake effect snow. My brother, who has lived in this area for many years, warned me about lake effect snow: There's a lot of it. More snow falls along the eastern side of Lake Michigan than falls even a short distance inland, he said. Later, I learned why.

Snow is caused by warm air rising. When cold Canadian air moves across the large body of a Great Lake's water, which is considerably warmer than the air, it is warmed from beneath by the lake. The warmed air rises. When it reaches the saturation point, this unusual amount of rising warm air can result in an unusual amount of snow. The infamous "snow belt" along the shores of Lake Erie and Lake Ontario—including Buffalo, New York, for example—can be nearly buried in lake effect snow.

Lake effect snow isn't ordinary snow, either. It can be much lighter. Generally, it takes about ten inches of inland snow to equal one inch of precipitation (water), but it may require thirty or forty inches of lake effect snow to equal this amount. Lake effect snow usually falls during very cold temperatures, because cold air can't hold much moisture. This is why clearing snow here sometimes feels like shoveling feathers.

Today I caught some falling snow on a piece of black paper. Even

---

SNOWFLAKE OR SNOW CRYSTAL?

A snow *crystal* is an individual ice crystal; a snow *flake* is many crystals stuck together. You can get both when it snows: individual crystals and big cottony flakes.

---

without a magnifying glass, I could see the star shapes I have always thought of as snow. I've discovered, however, to my surprise, that these make up only about a fifth of the snow that usually falls in most areas. There are other kinds, too.

❧ ❧ ❧

THE HEXAGONS OF SNOW

Six is the magic number for snow; a snow crystal is always hexagonal. Some snow crystals are the beautiful stellar (star-shaped) crystals you probably think of as snowflakes, but a snowstorm may contain several kinds of snow crystals. Some of these crystals can combine, for even more complicated shapes:

Plates:	Hexagonal, plate-shaped crystals
Stellar crystals:	Intricate, hexagonal, star-shaped crystals
Hexagonal columns:	Usually transparent, six-sided ice columns
Ice needles:	Slim, hexagonally sided rods
Asymmetrical crystals:	Haphazard, plate-like crystals
Snow pellets:	Rime-covered, hexagonal crystals; like tiny snowballs

JANUARY 3

Blue Jay

Today the sun came out for a couple of hours, and I walked carefully down the quiet, icy little road along the lakeshore, past old frame cottages, looking for something I recognized under the snow. Just standing in one place I could probably see fifty evergreens I didn't know. I certainly didn't want to start with those. I also saw quite a few scrambling, black- or gray-colored squirrels. I didn't know what those were, either.

Then a bird flew by. It was a crow. I even knew its call: *Caw! Caw!* Everyone knows crows. I saw a cardinal, too, passion red against the snow. Then a Blue Jay screamed. In ten minutes, I'd found three big, bright, noisy birds that I recognized. My decision was made: I would begin with birds.

Of these three commonly recognized birds, my favorite is the Blue Jay. Many people seem to find the Blue Jay ill-mannered, raiding the bird feeder and chasing off smaller birds. I like it anyway. If the Blue Jay weren't so common, it would inspire gasps of admiration for its relatively large size, intense blue color and perky crest. I find I share many characteristics with this colorful, bold, loudmouthed bird.

੨ਥ ੨ਥ ੨ਥ

WHERE THE JAYS ARE

In the East, a large, blue, crested bird is probably a Blue Jay. In the West, it is either a Stellar's Jay or the duller Gray Jay. The Belted Kingfisher, which is also blue, crested and white-breasted, does resemble a Blue Jay, but it's larger, with a dangerous-looking beak and a wide, blue band across its chest. (See page 98.)

FINDING A STARTING PLACE

Starting with something you know makes nature less intimidating—and even more astonishing. Your own knowledge, time of year and location will tell you where to begin.

JANUARY 4

Bird Feeders

It's been getting too cold to go to the birds, so I decided to bring them to me. What I needed was a bird feeder, the bigger the better. After looking over the selection at the local hardware store, I chose a big forest-green metal contraption with a vertical circular screen, through which birds could pick out one black sunflower seed at a time. It had a nice, cone-shaped roof and four wooden perches, and it held nearly three pounds of seed.

"I have one of these and the birds are all over it all day!" assured the affable owner. It was expensive, but I bought it, along with five pounds of black sunflower seeds—I'd never seen black ones before—which smelled rich and oily, were smaller than the big striped ones I was familiar with, and cost more.

The deck of my apartment extends from the second floor of the house, overlooking the lakeshore. My new feeder soon hung heavily from one of the big hooks that had been conveniently installed in the deck roof, for summer geraniums. Nearby tree branches provided handy perches for birds to check out the dining room before flying in for a feed. The waist-high enclosing wall was just wide enough for sprinkling a few extra seeds as an additional enticement. It looked irresistible.

BEFORE YOU BUY A FANCY BIRD FEEDER. . . .

The bird feeder I bought was my first mistake. What worked—and didn't work—for me may not be the same for you. Most libraries and bookstores offer many good books on attracting backyard birds, but every house and neighborhood has its own problems. Find a neighbor with a lot of bird feeders and introduce yourself. Or, as I did, find the best one by trial and error.

JANUARY 6

Gray Squirrel

My magnificent feeder was ignored for a day. Then yesterday, a big black squirrel leapt lightly from the balcony wall, wrapped its body around the feeder and stayed there for several hours, patiently picking the seeds out one by one. "Well," I thought, "it's awfully cute. The seeds aren't going down too fast. Why bother him? Or her?"

I had intended beginning with birds, but the squirrel had beaten them to it. I went to the public library and looked up black squirrels. Not one book pictured black squirrels. Nor were black squirrels listed in the index. Were black squirrels rare? They certainly weren't rare around here. The little rascals were everywhere, zipping up and down trees, pounding onto my deck from nearby branches, bounding across the road with that scalloped squirrel lope.

I wondered if there were so many species of mammals that something as common as a black squirrel wouldn't be included in the guidebooks. How inclusive were these books, anyway? (See page 313.) By the end of the day, I still didn't know what kind of squirrel was robbing my bird feeder. This was not an encouraging start for my year-long project. I continued to call my squirrel—mistakenly as it turns out—a black squirrel.

WHAT IF YOU GET THE NAME WRONG?

Don't worry about it. Perfection is not the point here. If you stick with this nature business for even a few months, eventually the "right" (generally accepted) name, along with many other bits of nature data, will fly at you as if you've been magnetized! In my opinion, it's the process of naming nature that's transforming; accuracy is secondary.

๕ ๕ ๕

WHEN IS A BLACK SQUIRREL A GRAY SQUIRREL?

It wasn't until April that I discovered that black squirrels are actually Gray Squirrels. Some species of wildlife can vary in color, and these variations are called either a *morph* or a *phase*, depending on which guidebook you're reading. (The old term *phase* is misleading, because the color is permanent—not outgrown—so some authors have replaced it with *morph*.) Like kittens, gray- and black-colored Gray Squirrels can come from the same mother.

JANUARY 8

Guidebooks

The only bird visiting my feeder in the past two days was a small, black-throated thing, but I had no idea what it was. It's no wonder more birds didn't come: They'd have had to fight off the six black squirrels that pounded around on my deck yesterday, leaping on and off the feeder, on and off the deck walls, thudding from branch to boards. Sometimes there were two on the feeder at once. The noise was incredible and it went on all day. I couldn't chase them off: I'd taped the deck door shut to stop the near-zero winds from screaming into the living room. I finally untaped the door and stuck four unwound coat hangers through the wire feeder holes. Eight daggers now stuck out in all directions, rattling dangerously. It took the squirrels two hours to figure it out.

So I had two problems. First, what was that black-throated bird? Second, how could I feed it out of reach of the squirrels? I solved both problems at the bookstore. The first solution was easy: A display of little plastic bird feeders met me at the door. They weren't very big, but they could be stuck on a window—out of a squirrel's reach—with a plastic suction cup. I bought one immediately.

Choosing a bird guide, however, turned out to be a real can of worms. Who'd have thought there'd be so many guides to birds? There were Audubon, Peterson's, Simon & Schuster and Golden guides, to name a few. There were fat guides and smaller guides for beginners. There were guides to local birds, guides to national birds, guides to eastern U.S. birds, guides to western U.S. birds, guides to world birds. Some used photographs, some were illustrated.

I was baffled. Was one guide about the same as another? How could I find out? The bookstore clerk was little help. Well, I had to buy *something*. "When in doubt," I told myself, peering at pages for more than an hour, "go with what you know." The name I knew was Audubon; I bought the *Audubon Field Guide to North American Birds (Eastern Region)*. (I also bought a mammal, a wildflower and a tree guide.) I was sorry I'd bought that particular Audubon guide. Although I'm glad to have it now—it contains a great deal of interesting information—I found it awkward to use for field identification. If you need help in choosing a guide, you'll find some recommendations in "A Guide to Guides," page 313.

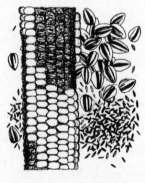

JANUARY 9

Bird Seed

When I got home yesterday, I put a cup of black sunflower seeds in my little plastic feeder and stuck it up on a window. No squirrel even got close to this one!

I left the big bird feeder up, hoping it would attract some birds, but all I've done is fight squirrels. Maybe the birds can't smell the food in the cold, I worried. Backyard bird books that I got from the library mentioned that it can take birds weeks to find a feeder, but what if they never came? Maybe I had the wrong food. Maybe some variety would help. I bought softball-sized, string-netted, seed-and-suet "bird balls," as they were labeled by our local grocery. I bought a wire suet feeder and packed it with seeded suet peanut-butter blocks. Soon five hunks swung from the deck overhang. I also filled my new plastic feeder from a five-pound bag of "Wild Bird Food." Still nothing.

ða ða ða

BIRD SEED: WHO EATS WHAT

Before you buy a cheap bird food mix, check the label: Is one of the following seeds or grains listed among the first (therefore main) ingredients? Seeds that, according to a rather thorough study on the subject,* birds rarely touch are wheat, milo, red millet, white rice, hemp, buckwheat and hulled whole oats. Here are some popular menu items:

Canary seed: finches, Common Redpoll, chickadee, many sparrows, Tufted Titmouse, Rufous-sided Towhee

Cracked corn: Blue Jay, cardinal, Mourning Dove, grosbeaks, Common Redpoll, many sparrows, Rufous-sided Towhee

Millet (pearl, small golden, white, or proso): Cardinal, Mourning dove, Common Redpoll, many sparrows, Rufous-sided Towhee, Carolina Wren

Peanut hearts: Blue Jay, chickadees, finches, Pine Siskin, grosbeaks, nuthatches, Tufted Titmouse, several woodpeckers, Carolina Wren

Suet: woodpeckers, chickadee, flicker, nuthatches, Tufted Titmouse, Carolina Wren

Sunflower seeds: Blue Jay, cardinal, Crossbill, chickadee, Mourning Dove (hulled), many finches, Pine Siskin, nuthatches, many sparrows, Tufted Titmouse, Rufous-sided Towhee, several woodpeckers

Thistle seeds (black): finches (they *love* this seed), Pine Siskin, Mourning Dove

* By Richard A. Viggars, of the Pennyfeather Corporation in Greenville, Delaware.

JANUARY 10

Fox Squirrel

Today I watched four black Gray Squirrels corner a bigger, rusty-colored squirrel on a skinny branch near the top of a tree, all five tails curling north in the stiff wind. Three black squirrels were positioned on the trunk and one at the end of the branch, trapping the bigger squirrel in between. All contestants chattered aggressively. The big squirrel squirmed for five minutes and then suddenly dashed at the three on the trunk, who fled instantly. Using my new mammal guide, I soon identified the victor as a Fox Squirrel. As it turns out, Fox and Gray Squirrels are the only two large tree squirrels in the eastern United States, which simplifies things. Still, how does one tell the two apart?.

The Fox Squirrel isn't nearly as ubiquitous here as the Gray Squirrel, which seems to have nearly no sense of territory. The Gray Squirrel shares with just about anybody, not roaming much more than two hundred yards from home. The Fox Squirrel will chase other Fox Squirrels from its larger range—up to several acres—but it will tolerate the Gray Squirrel. The arrangement usually works well, today having been an exception: Gray Squirrels are early risers, napping in the middle of the day, while Fox Squirrels sleep in, rising when their competition is napping. If today's skirmish was any indication, one Fox Squirrel can easily take on four Gray Squirrels, which love to quarrel, but rarely actually fight.

🙠 🙠 🙠

GRAY OR FOX SQUIRREL?

A black squirrel is a morph of the Gray Squirrel; a rusty-tinged gray squirrel is usually a Fox Squirrel; but a gray-colored squirrel can be either one. The usually rusty-tinged Fox Squirrel can have a gray morph in some eastern states, or a gray morph with a black mask in some southern states. Both of these morphs have white-fringed tails. The full-grown Fox Squirrel is bigger: about three pounds and two feet from nose tip to tail. The Gray Squirrel weighs in at two pounds and measures twenty inches.

NOT MANY BIG TREE SQUIRRELS

With a few exceptions, there aren't more than two kinds of big tree squirrels in any one area of the country. Usually, a big squirrel is a Gray Squirrel, common throughout the eastern United States as the Eastern Gray Squirrel, and in the West Coast states as the slightly larger Western Gray Squirrel.

BIG TREE SQUIRREL	RANGE	DESCRIPTION
Eastern Gray Squirrel	East	Gray or black
Eastern Fox Squirrel	East (but not New England)	Rusty, gray or masked
Western Gray Squirrel	West Coast states	Gray
Arizona Gray Squirrel	Southwest	Gray
Tassel-earred Squirrel	Southwest	Gray, tufted ears, white, or white-lined tail

JANUARY 11

Red Squirrel

I still don't have any birds at my feeders, but a new squirrel came on deck today: a chipmunk-sized, reddish squirrel with a skimpy-looking tail that it held over its back, in the manner of squirrels. A friend had warned me about these little rascals, calling them "Pine Squirrels." "If they get in your attic, you're in real trouble," she said. Well, there was no attic here, and this squirrel was really cute, and much shyer than the black ones. It took nothing to chase it away, so I didn't try. Anyway, there was only one.

There wasn't likely to be another. Although shy with humans, this was a Red Squirrel, which is fiercely territorial, noisily keeping away all others of its own species. It's sometimes called a Pine or a Spruce Squirrel because it prefers evergreens, feeding on pine and spruce nuts instead of the hardwood nuts loved by the Gray and Fox Squirrels. Red Squirrels are known for the tunnels they make under the snow as they look for winter food.

☙ ☙ ☙

JUST TWO SMALL TREE SQUIRRELS

The Red Squirrel is the only small tree squirrel found in the northeastern quarter of the United States, as well as throughout the Rockies; there are no small tree squirrels in the Southeast or Texas. Its West Coast near look-alike is called the Chickaree. Since their ranges don't overlap, identification is a snap!

JANUARY 13

Black-capped Chickadee

I was awakened this morning by wind gusting off the lake, rattling the windows. A dazzling, spotlight moon moved through the thick morning dark, between the swaying black silhouettes of trees. The sky became bluer and pink clouds began puffing around the moon before it slipped into a strip of inky water beyond the great white shelf of ice. By the time I'd made coffee, huge clouds, flesh-colored on top, light gray on the bottom, were sailing majestically toward me over the lake.

With them came the birds! There were several kinds at my feeder this morning, and they all looked alike to me, except the little black-throated one, which, after much paging through my new bird guide, I identified as a Black-capped Chickadee. The chickadee is a small, black, beige-breasted and gray-winged bird, jolly and bouncy. It dribbles through the air like a little basketball, lighting on feeders or hanging upside down from the suet balls. When I went out to fill the small feeder, a chickadee lit on the larger one, an arm's length away. I was amazed that a bird would be so bold.

ᴥ ᴥ ᴥ

HOW TO TELL A CHICKADEE

A small (4½ inches)* bird with a black cap, throat and beak and white face is probably one of seven species of chickadee, all a close variation on the Black-capped Chickadee, but more than two rarely show up in any one range. (See also page 290.)

CANADA; NORTHERN TWO-THIRDS OF THE U.S.
Black-capped Chickadee
Boreal Chickadee
Siberian Tit (Alaska and Canadian tundra)

WESTERN U.S.
Black-capped Chickadee
Mountain Chickadee

* All bird measurements in this book are a beak-to-tail-tip measure, unless indicated otherwise.

CHICKADEE·DEE·DEE!

Still not sure? Listen. Chickadees chatter. If you don't know their name, they'll announce themselves in a hoarse, double-noted *Chick-a-dee! Chick-a-dee-dee-dee! Chick-a-dee!*

SOUTHEASTERN U.S.
Black-capped Chickadee
Carolina Chickadee

PACIFIC COAST
Black-capped Chickadee
Chestnut-backed Chickadee

SOUTHWESTERN U.S.
Mexican Chickadee

JANUARY 15

White-breasted Nuthatch

This morning, when a White-breasted Nuthatch flew in—blew in?—I thought I was seeing a rare bird, its markings were so clean and beautiful. It looked dressed for a banquet. The White-breasted Nuthatch has a black cap (the female's cap is blue-gray) like the chickadee, bluish wings and tail and a white breast, but I'd never mistake a nuthatch for a chickadee. It has totally different body language. Maybe it's the way a nuthatch perches, its head stretched forward expectantly. Or hanging upside down, cocks its head up? Or maybe it's the sleek build, the swing of its hood into a neat, formal half-collar, or the way its stubby tail vanishes under folded wings.

The White-Breasted Nuthatch, found nearly everywhere in the United States and Canada, is a popular feeder bird, loving sunflower seeds and suet. It gets its name from its habit of inserting a seed or nut into a crevice of bark and using its beak to hammer it open. Its main diet, however, consists of insects, eggs and other delicacies it pries from the bark of trees while trotting up and down the trunks or hanging upside down from larger branches.

I've noticed that several kinds of birds seem to hang out together, flying in every several hours for another feed, including nuthatches and chickadees, and I'm beginning to know when they're here: I just listen for the White-breasted Nuthatch's toy trumpet call: *Yank! Yank! Yank!*

༄ ༄ ༄

THE UPSIDE-DOWN BIRD

Nuthatches are the only birds that can walk headfirst *down* a tree trunk. They can walk up, too, of course. Four nuthatches frequent North America. The tiny Red-breasted Nuthatch shares most of the White-breasted Nuthatch's range: most of the United States and southern Canada. In the West, look for the Pygmy Nuthatch as well, and in the Southeast, the Brown-headed Nuthatch. You'll need a guidebook to tell most nuthatches apart.

JANUARY 18

Herring Gull

Today I took to the beach to enjoy a January thaw. Last week the lake had been frozen so far out that I couldn't hear the surf; it was a strange white world down there, lumpy and silent. Now I could walk on sand at the water's edge, hopping on melting ice piles to avoid the bigger waves. Crystalline ice-sponges sparkled on the wet sand. Icicles dripped daintily from cave-like overhangs. Sunlight spangled everything.

After walking half an hour, I noticed a gull on a promontory of rock and fallen birches. I wondered why it didn't fly away as I approached, until I saw that it trailed a mass of almost invisible fishing line, which had tightened around the gull's neck and tangled in a wing: Every time it tried to fly, it choked itself. Catching the bird easily, I held it against my jacket and began walking quickly back. The gull was soft and warm, and it lay so still I thought it had died of fright. Now and then, though, it reassured me by blinking its round, black eyes. When I came to a friend's beach, I climbed the nearly one hundred steps of steep stairway that scaled the towering bank. She wasn't home, but I saw a police car approaching, and I stopped it.

The startled officer remembered his first aid kit and found me a bandage scissors. He held the bird while, bare-handed and trembling, I dug for the line under the neck feathers. The gull submitted docilely. The feathers were surprisingly deep and downy, the neck beneath as thin as a pencil, and the cord was so taut I had trouble getting the scissors under it.

Finally, I snipped. The bird shook its head and almost escaped, flapping hard, but I still had to get the line out of its wings. I stretched out the right wing, marveling at the layers of feathers, avoiding the bill that now banged viciously at my hands as I cut out several steel sinkers and a mass of tangled line.

When we set the bird down, the gull began walking straight up the middle of the narrow road that was lined with woods and cottages. I shooed it toward the lake. It tried its wings, then flapped harder. Suddenly, perhaps realizing that it could fly without choking, without pain, the gull flew away. I'd read that the Great Lakes and other bodies of water are full of such bird- and fish-killing debris. It saddened me to see a bird tortured so unnecessarily.

At home, all the gulls in my bird guide looked alike—almost two dozen species were listed for the United States alone. I was reminded of early childhood puzzles: Which Picture Is Different? I found, however, that when I eliminated species not found around the Great Lakes area, I was dealing only with six. When I eliminated rarely seen species, I was down to three: Herring Gull, Ring-billed Gull and Bonaparte's Gull. Although all three have gray backs and white bodies, these were not hard to tell apart: The Herring Gull has a plain yellow bill; the Ring-billed Gull's bill is ringed with black; and except in winter, the Bonaparte's Gull's bill is black (in summer, its whole head is black). My gull was big, with a gray back, white body, yellow bill and pink legs. It had to be a Herring Gull.

Thank goodness my gull wasn't brown. Gulls usually don't reach adult plumage until their third winter, and until that time, most are more or less brown. Trying to identify adolescent gulls might make me give up birding for good! I'm also glad I don't live on the Pacific Coast, where the choice would have been more difficult.

❧ ❧ ❧

HERRING GULL LOOK-ALIKES

If you want to be a West Coast gull expert, you'll spend a lot of time with your bird guide: All these Herring Gull look-alikes, as well as other gulls, are found there. Most of these gulls have yellow bills and black-tipped gray backs. One difference seems to be the leg color.

GULL	RANGE	LEGS	OTHER
Herring	North America	Pink	
Ring-billed		Yellow	Ringed bill
Thayer's	West Coast	Pink	
California		Yellow	
Mew Gull		Yellow	
Heerman's		Black	Black tail
Black-legged Kittiwake	East and West Coasts	Black	

JANUARY 19

Northern Cardinal

"If it weren't for cardinals, I don't know how I'd get through the winter!" Deb, the director of our village library, told me today. We were discussing winter birds as I checked out my books. It wasn't just the cardinal's jazzy carmine color that she loved. It was the male's song, the second note of which rises cheerfully: *Birdy birdy birdy birdy birdy*. "You start to think there's nothing alive in the entire world, and this big red bird flashes past and whistles its heart out clear as Christmas and loud enough

---

CAN'T MISS A CARDINAL

The male cardinal is the only all-red, crested bird in North America, but you'll only see it in the eastern half of the United States. The female is yellow-brown with a red beak, as are the young. Almost as big as a Blue Jay, an adult cardinal measures about nine inches.

---

to wake the dead." That's not the only cardinal song, either: Researchers have recorded twenty-eight others.

I've been surprised at the large numbers of cardinals here. It used to be a southeastern bird; a hundred years ago, no one ever saw a cardinal this far north. It doesn't migrate, but seems to adapt well to inhabited areas. Cincinnati is crazy with cardinals, for many years topping the Christmas Bird Count* with cardinals. (Cincinnati residents buy over four hundred tons of sunflower seed, the cardinal's favorite food, a year.) Now we have them here, lots of them, and they have spread throughout most of the eastern half of the United States, sometimes as far as southern Canada. It's certainly nice to get a little good news, in the face of so many vanishing species.

Cardinals haven't appeared at my bird feeders yet, so I am quite jealous of my friend Mary down the lake, who has at least three pairs at hers. She even showed me a cardinal's nest built in a shrub just outside her window. A cardinal's nest is easier to find than most nests, built as it is fairly near the ground, usually within eight feet.

Afterword: Later in the spring, I did get a pair of cardinals at my house. It being mating season, they acted like lovebirds. Often, the male would reach over and give his mate's beak a little peck. Or he'd crack a sunflower seed on the deck balcony and put the kernel in her mouth. Cardinals are good parents, too. They are so devoted to feeding their young that if the brood is lost, they sometimes feed the young of other species.

---

* A National Bird Count sponsored by the U.S. Fish and Wildlife Department and the National Audubon Society.

JANUARY 21

White-tailed Deer

Today I heard a deer bark. It was sunny and warm, a good day to explore the Saugatuck Dunes State Park a few miles north of me, a maze of fourteen miles of ski trails, oddly situated next door to a state prison facility.* Most of the snow had melted, so I had the rolling forest to myself. I was about halfway along a modest three-mile loop through lakeside woods when I came upon three deer browsing on the low evergreen branches. One of them made a sharp whirring sound; then all three wheeled and drummed a crashing retreat, their white-lined tails flipped up, hooves thunking on the needle-thick ground. I've always thought of deer as silent creatures, but I've since read that when startled, they sometimes sound an alarm.

I find it astonishing that our largest wild mammal (we have no bears here) need not be feared and is so full of grace. This is not to say that deer are always welcome. Deer are not as shy as they seem, invading yards in suburbs and even towns with the boldness of cats. A friend finally deterred her lovely marauders (without harming them) by stringing an electric fence around her garden, following Robert Frost's adage that "Good fences make good neighbors." This advice, she has come to believe, is not as cranky as it sounds.

Deer have not always been so abundant. Around 1900, unrestricted hunting and destruction of habitat had made deer a rare sight in many midwestern and some eastern states. By the 1920s, hunting restrictions had been introduced, including "bag" limits, protection of does and fawns and a restricted season. Despite their delicate appearance, deer are adaptable creatures. Although basically a browser of twigs, shrubs, grasses and herbs, they will also eat, among other things, acorns, fruit and bark. I find it amazing that any of them escape the dangers in their environment; hunters, highway accidents, cold, starvation, parasites and diseases and other disasters kill hundreds of thousands of deer every year.

---

* This prison has since been closed.

> ### THE WHITE-TAIL COLOR CHANGE
>
> In summer, a White-tailed Deer sheds its blue-gray winter coat of hollow hairs, which gives it grand insulation, for a cooler, red-brown coat.

I wasn't sure what kind of deer I had seen until later, when I discovered that I only had one choice.

Rutting season being over—it lasts from about September through December in the North, longer in the South—the males have dropped their antlers (males grow a new set of antlers every year), so I didn't know if the deer I had seen were male or female. They were probably all one or the other: Adult bucks and does, with their fawns and yearlings, tend to stick to groups of their own sex in winter.

ะ ะ ะ

ONLY TWO COMMON NORTH AMERICAN DEER

The White-tailed Deer is the only species of deer common to the eastern half of North America.* The West has two: the White-tailed Deer and the Mule Deer, also known as the Black-tailed Deer. To tell the two apart, look at the end a deer is quickest to show you: White-tailed Deer flip their white-lined tails straight up when they run, which is called "flagging." Mule Deer tails, which are black on top or on the end, are not raised in flight.

* Isolated pockets of unusual deer, such as the diminutive Pigmy Deer and Toy Deer, can be found in a few protected areas.

JANUARY 22

Tufted Titmouse

A transatlantic postcard from a friend visiting Bavaria informed me this morning that the German expression for "Are you nuts?" is *"Du hast wohl 'ne Meise?"* which translates literally as "Have you a titmouse in your head?" I found this information serendipitous, since I had begun reading about titmice only hours before checking my mailbox. The titmouse family pictured on the German postcard looked more like little crestless Blue Jays than our crested gray titmouse. However, if the German titmouse has the same persistently, unbelievably loud call as ours—*Peter, Peter, Peter, Peter, Peter!* (repeated four to eight times)—the expression rings true.

The titmouse that comes to my feeder hangs out with the chickadee/nuthatch crowd. It has a white breast, peachy flanks and darker gray upperparts, a perky gray crest (in parts of Texas, the crest is black) and big, round, black eyes. There is something cutely mousy about a titmouse; when it comes to the feeder, it squeaks. It's shyer than chickadees and nuthatches, which will fly greedily over my shoulder to attack a topped-off feeder. Not so the titmouse. It sticks tight on a nearby branch, its shoe-button eye staring and wary.

The titmouse is an easy bird to identify: Nothing else looks quite like it. A few gray flycatchers have slightly pointed heads, but you're unlikely to see them at your feeder, since, true to their name, they don't eat seeds and rarely hang around cold regions in winter.

❧ ❧ ❧

WHICH TITMOUSE?

A small (4½ to 5½ inches) crested bird with gray uppers and a white breast with buffy flanks is likely one of three titmice that strongly resemble each other, but rarely overlap ranges: the Tufted Titmouse in the eastern half of the United States, the Plain Titmouse in the southwestern quarter and the zebra-faced Bridled Titmouse in a small area along the Mexican border.

JANUARY 24

Downy Woodpecker

I was very happy today that no neighboring cats have yet appeared to drool and chatter helplessly beneath the hunk of suet that hangs over the deck, for a lovely little woodpecker came looping through the fog, lit there and, feeling quite safe, stayed long enough for me to identify it. It was a Downy Woodpecker, the most common one in the East, and smallest of all North American woodpeckers. Initially, I confused it with the Hairy Woodpecker, which is also common here. Only after seeing both frequent my place for a week am I getting a sense of who's who.

ۿ   ۿ   ۿ

DOWNY AND HAIRY

A small, striped-winged, black-and-white woodpecker with a red patch on the back of its head is likely either a Downy (5 to 6 inches) or a Hairy Woodpecker (7 to 8 inches). Both are white-breasted and -backed, with black-and-white tail and head. To tell them apart, look at the beak: The Downy's narrow, short beak looks fairly harmless, but the Hairy's beak is a heavier, no-nonsense one, about three times bigger. Both birds are common throughout North America.

WOODPECKERS: SPECIAL FEATURES

Woodpeckers have unusually thick skulls to keep their tiny brains unscrambled during the tremendous speed and force of their persistent pecking. The Downy, although small-sized and -billed, is one of the hardest-headed woodpeckers, favoring, as it does, hardwood trees as its source of bugs and larvae. A woodpecker's tongue is unusual, too. Often barbed at the end, it's too long to be stored in its mouth. Rooted in one nostril, it curls over the top of the skull before extending through a hole in the bottom beak.

JANUARY 25

Common Crow

Having at last admitted defeat, I removed the large bird feeder and made my peace with the squirrels. I even feed them now, dribbling seeds along the top of the deck wall. At 10:20 this morning, four black squirrels were nibbling their way through breakfast in rare harmony, quarreling only as they bumped into each other, when four black crows cruised in and landed on a nearby branch, cawing loudly. A noisy skirmish ensued, crows vs. squirrels, as black fur and feathers flew. After an easy victory, two of the crows pranced arrogantly back and forth on the deck wall, preening but wary. I thought them quite as magnificent as they seemed to think they were, and enormous, too. As soon as I moved across the room, however, they flew off, cawing bitterly. The squirrels quickly resumed their positions, winning the war if not the battle.

Around here, Common Crows hang out in raucous gangs of four to eight birds. Despite the air of stupidity cast by their thuglike behavior, crows have been proven to be one of the most intelligent of all birds. It's mostly in fall and winter that crows gang up like this, at night roosting in huge crowds. In spring and summer, they quiet down, become family birds and claim home territories. Crows even are said to mate for life, although I've read that many birds once thought to mate for life actually hang in there only a few years.

HOW TO FIND AN OWL IN DAYTIME

If you hear crows or Blue Jays making a particularly loud racket, look for an owl, which is easier to see in daylight than after dark. Crows, as well as Blue Jays, are known for *mobbing*—attacking a larger bird in gangs—and harbor a special grudge against hawks and owls. Crows hate owls so much (the Great Horned Owl especially finds them quite delicious) that researchers sometimes accompany a stuffed owl with hoots to attract crows.

☙ ☙ ☙

BIG BLACK BIRDS

The Common Crow (17 inches), abundant throughout North America, is the only all-black bird in the middle portion of the United States. Elsewhere there are several others that look so similar that the best way to tell them apart is by their calls, a lot to ask of an amateur. The Common Raven (21 inches) croaks, and is found mostly in the West and far North. The Fish Crow (Southeast and East Coast) croaks like a frog, the Northwestern Crow (far Northwest) is hoarser than the Common Crow, and the Chihuahuan Raven (Southwest) sounds much like the Common Raven.

JANUARY 27

Yellow Perch

This afternoon I drove north along the shore of Lake Michigan, an adventure not as straightforward as one might think. Every twenty or thirty miles, I ran into a river that rushed toward the lake, bumped into the sand dunes and ballooned into a small lake. From each of these little lakes a channel had been dug to the big lake, although no bridge was provided, since heavy boat traffic would require an expensive and inconvenient drawbridge. So I went around. And around. And around. It was at the fourth balloon—White Lake at Montague—that I came upon a scene worthy of Brueghel. Under a thousand sunlit, wheeling gulls lay a white, frozen lake littered with red, turquoise, green and gray shanties and perhaps a hundred hunched forms of ice fishermen. I decided it was time I named something.

Walking on water, whatever its molecular arrangement, scares me silly, but I made my way gamely toward the nearest pair of fishermen, a redheaded man and his teenage son, who welcomed me warmly. Talking, they said, didn't bother the fishing none; in fact, the company's what they come out here for. They came well equipped: A makeshift sled held a leg-long shiny red drill topped by a small motor; an electronic fishfinder, across the screen of which blipped little gray goldfish; black plastic cases for the foot-and-a-half-long ice-fishing poles they steadied on their knees over holes in the ice; a couple of buckets; a thermos of coffee and assorted other gear. I was amazed. I had assumed that ice fishing would be more—well, primitive.

I appeared to be the only woman on the lake, an instant curiosity. As we talked we were joined by other fishermen, each of whom settled down, augered a nice big hole through the eight-inch ice, dropped in a line and joined the conversation. The Swiss cheese effect inspired me to ask if anyone there had ever fallen through the ice. Every man told his own personal falling-through-the-ice story, which he found hilarious. One large man had been following some fishing buddies back to shore when he suddenly disappeared, going straight through ice the others had just traversed. His friends, highly amused, rescued him easily and have been ribbing him ever since. Another man had been fishing too close to the channel connecting White Lake to Lake Michigan, lured by fish schools as thick as the ice there was thin. He'd gone down, was rescued, but had to have his shack hauled out later, lest he be fined. (Coincidentally, on my way home later, I heard on the radio that another man had just fallen through the ice on a nearby lake.) Falling through the ice seemed to these good-natured sportsmen a marvelous probability.

They were catching Yellow Perch, small yellow-bellied fish that are usually six to eight inches long, although they can get bigger. There's no size limit on Yellow Perch in Michigan, and no season, either. Each licensed person can legally take home a hundred Yellow Perch on any day of the year. Buckets flopped with twenty or thirty hand-sized fish each, but it wasn't a good day. "Yesterday my boy and I took home 145 of 'em," said my host. "I was up until three in the morning cleaning them." I asked about what bait was best, and after a lively discussion, they decided it was fish eyes. "My seven-year-old son can pop 'em right out," said a man in an orange-and-olive camouflage suit. "Makes you wonder if fish don't go swimming around down there, trying to eat each other's eyes out."

As I left, the sun was going down behind a fringe of trees, trimming the ice with black lace. The gulls, turning suddenly, flashed in the late light like stars.

ə ə ə

GREAT LAKES PERCH

There are two common Great Lakes fish from the Perch family: the Yellow Perch and the Walleye. Although the Walleye grows bigger, I never could tell them apart, because they are both fairly yellow and are ringed with wide black stripes. Count these: The Yellow Perch always has seven; the Walleye has more.

JANUARY 30

House Finch

Until now, the only feathered visitors that I had any real problem identifying were the Hairy and Downy Woodpeckers. Today, however, a pair of little finches—a brilliant red-headed and -breasted male and a streaky brown female— flew in and completely baffled me. Were these House or Purple Finches? House Finches were so common in California that I thought I recognized them. However, the House Finch has only recently expanded its range this far; ten years ago, it was rarely seen in Michigan.

The male of both species is red—the Purple Finch is not purple— with streaky brownish wings and tail. Guides seemed to indicate that the House Finch is very red, while the Purple Finch is very, *very* red, a comparison that didn't help much. Purple Finches prefer the woods and are shyer than House Finches, so my visitors were probably House Finches.

ə ə ə

LITTLE RED BIRDS

In most of the United States and Canada, a little red bird (5 to 6 inches) is probably a male House Finch or Purple Finch. Look at the breast: The male House Finch's is streaky, while the Purple Finch's is plainer. In the North, it might also be a Redpoll,

---

FEMALE "LBJS"

So far, female chickadees, nuthatches, titmice and woodpeckers have more or less resembled the males. Many species' females, however, like the finches, tend to be what my birder cousin calls "LBJs," or "Little Brown Jobs," resembling females of other species more than the males of their own species. As one committed to bringing more attention to the female of my own species, I hate to ignore difficult-to-differentiate female birds, but I don't see how a beginner can hope to deal with the subject. I reluctantly choose to focus primarily on the easier male birds, although many of those, especially sparrows, are called "LBJs" too.

---

which has a distinctive black patch under its beak. In the West, it might also be a Cassin's Finch, which closely resembles both the House and Purple Finch, or the much duller Rosy Finch.

HOW TO ATTRACT FINCHES

Thistle seed—long, black and narrow—is finch caviar, practically irresistible to the beautiful red finches as well as Goldfinches (see page 57). Although thistle seed is expensive, the colorful song-and-dance it attracts—finches are gregarious birds—makes it easily worth the price. Special finch feeders are available which dispense the seeds through holes so tiny that squirrels usually ignore them.

February

Evergreen Woodfern

I seem to have run out of new feeder birds to identify just as my health, which was suffering when I moved here, has recovered. Now I can foray into the bitter February lakeshore wind to explore the evergreens, those remarkable plants which somehow manage to look alive when everything else looks dead. Not all of these plants are bushes and trees: Today I was amazed to find an impressive patch of ferns, which I've always regarded as fragile and delicate, springing from the snow-covered dune that rises from my beach.

The few ferns that I've adopted in my life did not long survive my care. The fronds of this fern, however, were tough, leathery and nearly as long as my arm, radiating from a center root like the petals of a gigantic, otherworldly flower. They otherwise resembled those feathered greens that sometimes accompany a gift of red roses. I cut a frond—it was too tough to pick—and brought it inside. Just as I had feared, about twenty ferns in my fern guide appeared to look just like mine. To "see" ferns, I would have to learn some fern facts.

My fern was the twice-cut kind. I resumed hunting for it in my new fern guide (I am already beginning to accumulate guidebooks), finally finding it in a "Larger, Coarser Ferns" section: It was an Evergreen Woodfern, also known as a Marginal Woodfern because the spores lie along the underneath edge, or margin, of the leaves.

FIDDLEHEADS

Fiddleheads are not a kind of fern, but the new growth of any true fern, which uncurls from the bulbous root like the head of a violin. The fiddleheads of some ferns, like the Ostrich Fern, are edible—cut in the spring and boiled. Although no fiddleheads appear to be poisonous, many taste bitter, and some ferns are poisonous to livestock when mature. It's possible to confuse fiddleheads with the new growth of very poisonous wildflowers like Water Hemlock and Poison Hemlock.

ɘ ɘ ɘ

FAST FERN FACTS

1. The "root" the fronds grow from is a *rhizome*.
2. The tender new uncurling fronds are *fiddleheads*.
3. The dots under the leaves are *fruitdots*. Each dot contains many repro-
   ductive cells called *spores*.
4. The fern leaf is divided into four kinds:
   a. Once-cut: one big Christmas tree
   b. Twice-cut: a long stem with little Christmas trees
   c. Thrice-cut: a long stem with tiny Christmas trees growing from each
      side of its branches
   d. Other untypical forms

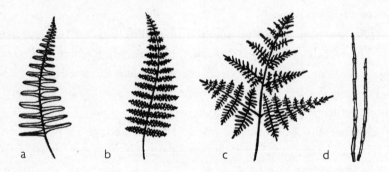

a   b   c   d

TWO COMMON KINDS OF WINTER FERN

The Evergreen Woodfern isn't the only common winter-green fern. There's also the
Christmas Fern, which, being a once-cut fern, can be easily told from the twice-cut
Evergreen Woodfern. Both are common in woodsy areas all over the country.

FEBRUARY 2

Eastern White Pine

Winter is a dandy time to start naming things outdoors: the amount of active plant and bird life isn't as overwhelming as in more verdant times. I'm not about to try to identify trees without leaves, so the only large green plant group left right now is evergreens, standing tall amid a wild, loose weave of brown and gray branches. I've recently heard several people call any needled evergreen a "pine." At a party a few nights ago my host boasted that he had over twenty different kinds of "pines" on his property, and I thought, I'll bet there aren't that many kinds of pines in the whole eastern United States. We were both right.

The most common evergreen around here is a graceful, long-needled pine, with soft needles that grow in bundles of five (sometimes four), so I decided to use it to start my arboreal education. What a fortuitous and encouraging choice it was. This pine turned out to be an Eastern White Pine, the simplest of all eastern evergreens to identify, and probably western, too.

ৰ ৰ ৰ

INSTANT WHITE PINE I.D.

Count the needles in each bundle: In the eastern United States, only the Eastern White Pine has four or five (4- to 6-inch) needles per bundle. In the West, a very similar tree is called the Western White Pine. The only other pine with long needles bundled in fives is the West Coast's enormous Sugar Pine, which boasts the longest cones (up to 26 inches) of any conifer. There are five other western pines with bundles of five *shorter* needles.

ARE ALL EVERGREENS "PINES"?

A cone-bearing tree with evergreen needles that grow in bundles of two, three or five is a pine tree. To confuse matters, however, pines are members of a larger group (family) which is also called pine, and which, in addition to pines, includes firs, Douglas Firs, spruces, hemlocks and larches, which, like pines, are also cone-bearing and needle-leaved, but don't grow needles in bundles. From here on, *pine* refers to a tree that grows needles in bundles, not the larger family.

THOSE CONES ARE FEMALE

Those woody "pine cones" that you see on evergreens are female cones. Conifers (cone-producing plants) produce both male and female cones, often (but not always) on the same tree. The male cones are usually on the bottom part of the tree, obvious only in early flowering and carrying huge amounts of pollen; conifers depend on the wind for pollination. Once pollinated, female cones can take one to several years to mature before opening to release their seeds.

FEBRUARY 4

Red Pine

The little road I live on follows the lakeshore south for miles, lined with cottages, but only a few minutes' walk to the north, it curves east, ending at the Kalamazoo River. Yesterday, walking toward the river, I found a stand of tall, straight, long-needled pines in someone's front yard. The long, dark, thick needles were bundled in groups of two. Today I walked south and found a similar stand of tall, stately pines. I thought the two groups of trees were the same species, but I didn't know which. They could have been Black Pines—imported trees called Austrian Pines in some guides—or Red Pines. Then I learned the trick for telling them apart and discovered, conveniently, that I had found one of each species.

ᕐᕐ ᕐᕐ ᕐᕐ

PINES: THE RED AND THE BLACK

Only in the Northeast are there pines that grow long (4- to 6-inch) needles bundled exclusively in pairs (none in threes). To tell them apart, bend back one of the needles: If it snaps easily, it's a Red Pine. If not, it's a Black, which, although not a native pine, is frequently planted.

FEBRUARY 6

Eastern Redcedar

Today being too cold to venture far, I decided to tackle the evergreen trees near my house. The closest, next to the front door in fact, was one of those scaly kinds of evergreens that I think of as "junipers." My tree guide, which covers all North American trees, at first confused me with over twenty species of scaly evergreens, but once I eliminated those residing happily in the western states, my portal tree was easily identified as an Eastern Redcedar. It was my lucky day: We who live in the eastern United States have only three scaly-type evergreens (even fewer in more southern states) to choose from.

Although the Eastern Redcedar is known for its distinctive aroma in "cedar" chests and closets, it's not a true cedar. It's a juniper, bearing berrylike "cones" as junipers do. The other scaling evergreens—cypresses and true cedars—grow small, woody cones. The only explanation I have found for this misnomer is that Eastern Redcedar smells like cedar. Redwoods are also sometimes called cedars for their aromatic wood. To confuse matters further, there is a Southern Redcedar (found along the Atlantic and Gulf Coasts, and in Florida), which closely resembles the Eastern Redcedar and is also a juniper, and a Western Redcedar (found in the northwestern states), which is a true cedar, with small flowerlike, woody cones.

Once I had identified an Eastern Redcedar, I began seeing them everywhere. Thriving in poor soil, the rusty-looking, flame-shaped trees often dot areas once de-treed, such as barren spots along our state highways.

WHEN IS A BERRY A CONE?

Junipers are officially *conifers*, or cone-bearing plants. So how do the scientists manage to class junipers as conifers? They call the berries "cones." All my guidebooks called these berries "cones," which at first confused me. A scaly-type evergreen with berrylike "cones," usually (but not always) dark blue, is not only a juniper, but a conifer as well.

৯ ৯ ৯

REDCEDAR GIVEAWAY

To quickly identify an Eastern Redcedar, look for bristly new growth. The Eastern Redcedar is the only scaly-type evergreen with two kinds of foliage: scaly, inner, older 0foliage and bristly new growth on the branch ends. In season, the dark blue "berries" help, too. Another tree with these two kinds of foliage is the Giant Sequoia, found mostly in limited patches of California.

FEBRUARY 8

Arbor Vitae

Today in a yard in downtown Saugatuck, I saw a row of flame-shaped shrubs much like Eastern Redcedar, but the scaly branches were a rich, almost spring green color, quite different from the duller, darker Redcedars. The flat foliage looked very much like that of the Northern Whitecedar, a native tree in my guide, but this was definitely a shrub. I could find it nowhere in shrub guides, however. It turned out to be one of the many varieties of Arbor Vitae (see page 37), but this one was a popular ornamental shrub. I have learned my lesson: Wandering into gardens to identify plants is wandering into a confusing world of imports and garden plants which often can't be found in guides to wild species. Around here, though, the two evergreen shrubs—Eastern Redcedar and Arbor Vitae—are seen so often that it's useful to tell them apart.

৯ ৯ ৯

HOW TO TELL EASTERN REDCEDAR FROM ARBOR VITAE

Both the popular Arbor Vitae shrub and the Eastern Redcedar tree are shaped like candle flames, with scalelike leaves. If the new growth is prickly and it has blue berries, it's an Eastern Redcedar. Arbor Vitae grows in flat, smooth, bright green sprays and has small (½-inch) woody cones.

WHERE THE SCALY-TYPE EVERGREENS LIVE

Junipers and cedars represent all the scaly-type evergreens found in the eastern United States. Scaly-type cypresses and three of the five most common cedars are found only in the West Coast states.* If you live in the Rocky Mountain or Southwest parts of the country, your big worry is trying to figure out all those junipers! Here is what to look for in each region:

ATLANTIC STATES	SOUTHEAST	NORTHEAST**
Atlantic Whitecedar	Atlantic Whitecedar	Northern Whitecedar
Eastern Redcedar	Eastern Redcedar	Eastern Redcedar
	Southern Redcedar	Common Juniper

ROCKY MOUNTAINS	SOUTHWEST	WEST COAST STATES
Rocky Mountain Juniper	Rocky Mountain Juniper	Western Juniper
One-seed Juniper	One-seed Juniper	Common Juniper
Common Juniper	Common Juniper	Incense-cedar
Utah Juniper	Utah Juniper	Western Redcedar
	Pinchot Juniper	Port-Orford-cedar
	Drooping Juniper	Alaska-cedar
	Alligator Juniper	Monterey Cypress
	Arizona Cypress	Gowen Cypress
		Macnab Cypress
		Tecate Cypress

* Baldcypresses, a different group, are found in the Southeast and have needles, not scales.

** "Northeast" is used in this book to refer to the northeastern quarter of the country and includes the Great Lakes area.

FEBRUARY 10

Common Juniper

Maybe it's the unseasonably warm weather, but the birds are beginning to sing. When I walked through a woods just a ten-minute hike from here, I heard crow calls crashing through the quiet, in addition to jays, doves, cardinals and titmice. The titmice sang loudly and incessantly:

PEter PEter PEter PEter PEter PETE! A pair of woodpeckers jackhammered on dead trees with different notes, as if the trunks were giant musical instruments.

I was looking for the Common Juniper, but it must not be too common here, although it presumably does like sand dunes, among other dry places. I couldn't find one, but since the Common Juniper is one shrub I do recognize, I've decided to write about it anyway. The Common Juniper is the only one of all the normally scaly evergreens—cedars, cypresses (except for Baldcypresses) and junipers—that isn't scaly. It's usually a shrub, rarely becoming a little tree. The Common Juniper grows extensively over the northern third of the East, the northern three-quarters of the West and most of Canada.

❧ ❧ ❧

A NON-SCALY JUNIPER

A sprawling evergreen shrub that bristles with little three-needle whorls instead of scales is likely a Common Juniper. Its dark blue berries (cones), covered in gray wax, tell you right away that it's a juniper. Short, pointed needles establish it as a Common Juniper.

FEBRUARY 11

Scotch Pine

Between the interstate bridge and the shoreline, about two miles east of the lake shore, the Kalamazoo River swells into two small connecting "lakes," which, on a map, look remarkably buxom, with a long eastern arm slithering inland through wetlands and a short western arm horseshoeing north toward Lake Michigan. Tucked into the cleavage on the south side of the river is the quiet little town of Douglas; on the other side,

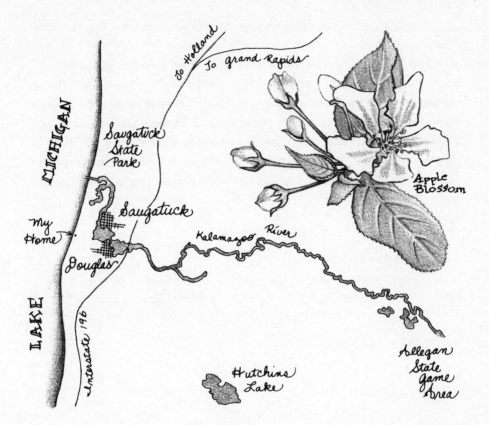

flanking the horseshoe arm, is Saugatuck, a lively town of tourist delights, such as delis, galleries and boutiques. I live a mile and a quarter due west of downtown Douglas and have noticed that the two towns seem to function socially, if not governmentally, as one.

I was driving around, trying to get my bearings in all this, when I came upon an unassuming little park wedged between the interstate and the first swelling of the river. The drive curving into the park was lined with crooked pines, like giant bonsais, with thick gray lower bark cracking to reveal sappy orange underneath, and very orange branches. I guessed them to be Red Pines, because the bark was so bright and the needles were bundled in pairs, but I was wrong. Red Pine needles, although also paired, are brittle and much longer, and Red Pine trunks grow straight and stately. I took a sample of needles home and identified the two-needle bundle right away as Scotch Pine. These needles were only two or three inches long, and slightly twisted, which is enough to identify this tree. But the bark clinched it.

The Scotch Pine isn't a native tree; it's a fast-growing, adaptable tree imported from Europe, the only conifer native to the British Isles. About a hundred years ago it was widely used for reforestation projects all over the United States, so one does come upon it in the wild. The variety of Scotch Pine used here grew charmingly but uneconomically crooked, so they are no longer used in forests. Bred into comely shapes, however, the Scotch Pine has become the most popular Christmas tree in America. (See page 298.)

🐦 🐦 🐦

ORANGE-BARKED PINE

A tree with red-orange bark on the branches and the upper part of the trunk—sometimes an outer gray bark cracks to reveal the inner orange—is probably a Scotch Pine. If it also has medium-long (1 ½- to 3-inch) needles bundled in pairs, your identification is secure.

FEBRUARY 14

Jack Pine

Today, cruising along country roads in my van, I felt like a real expert, spotting Eastern Red-cedars, White Pines, Scotch Pines and Red Pines at forty-five miles per hour. Then my eye caught a tree that looked odd. I stopped to check it out and found yet another two-needled pine. This one was a scraggly tree with very short needles bundled in little v's and cones, which clung to the branches in pairs, curving gracefully toward each other, like inverted commas. A Jack Pine was easy to spot once I thought of it as the "Pair Pine."*

---

* There are other North American pine species—the Shortleaf, Ponderosa and Slash Pines, for instance—with needles in bundles of twos and threes on the same tree.

The Jack Pine was an appropriately romantic tree to find on Valentine's Day, for the cone pairs on Jack Pines, each cone a mirror image of the other, can cling to the branches for years. Sealed with resin, the cones open only under great heat; Jack Pines are often among the first plants to grow back after a forest fire. Found in the Northeast and throughout Canada, the Jack Pine attracts many species of birds. There's a rare warbler— Kirkland's Warbler—that requires a Jack Pine habitat and breeds only in a small area of north-central Michigan.

❧ ❧ ❧

SORTING OUT TWO-NEEDLE PINES

The northeastern part of the country has more two-needle pines than any other area of the country: two trees with long needles (Red Pine and Black Pine), one with sharp, medium-long needles (Scotch Pine) and one with short needles (Jack Pine). Two of these, the Black Pine and the Scotch Pine, are imports. If you live elsewhere, you have fewer to choose from, although Scotch Pine may turn up almost anywhere.

WEST COAST	WEST	MIDDLE EASTERN STATES
Bishop Pine	Lodgepole Pine	Virginia Pine
		Table Mountain Pine

SOUTHEAST	NORTHEAST
Sand Pine	Red Pine
Spruce Pine	Black Pine
	Scotch Pine
	Jack Pine

FEBRUARY 15

Eastern Hemlock

On the north side of my house are nearly a dozen tall straight trees with soft, flat needles that seem to grow from their branchlets like hair combed from a middle part, although they actually grow in a spiral. Their foliage is lacy and delicate, soft to the touch. Rooted deep in the valley between two sand dunes, these trees are so tall that when I look out my second-floor windows, I see only their middles.

There are so many evergreens with soft, flat needles in the guide-books—numerous species of fir, yew, redwood and hemlock—that at first I was terribly confused. When I eliminated those out of range, however, the Eastern Hemlock turned out to be the only flat-needled conifer native to this area. (There's an imported Japanese Yew that is frequently used here as an ornamental shrub, but it is not a tree.) Even if there are other flat-needled trees in your area, the hemlock is still easy to identify.

Socrates, by the way, did not die from the juice of a hemlock tree. Hemlock trees, nutritious if not delicious, are not poisonous. There are species of wildflowers called hemlock, however, that most certainly are.

ɜ   ɜ   ɜ

HEMLOCK: A DISTINCTIVE TREE

The hemlock distinguishes itself by being the only flat-needled conifer to drip little (one-inch, or smaller) woody cones from the *ends* of its branchlets. (Yews have red berries; fir cones sit upright on the branch; redwood cones are also small but hang on little short-needled stems.) If you're still not sure, look at the underside of hemlock needles: Lighter green than the tops, they show two vertical white lines.

WHERE THE HEMLOCKS ARE

Most of the country, including the central, southern and East Coast (south of New England) states, have none of these North American hemlocks.

WEST COAST STATES TO ALASKA	NORTHEAST	APPALACHIAN STATES
Western Hemlock	Eastern Hemlock	Eastern Hemlock
Mountain Hemlock		Carolina Hemlock

FEBRUARY 17

Norway Spruce

I don't know why it took me so long to recognize a spruce. I pondered for the longest time over the huge pointed evergreen across the road from me. I'd stand on tiptoe, break off little branch ends that I could barely reach and then spend hours puzzling through my tree books. It drove me crazy: All the evergreens with short needles growing straight out of the branch looked alike to me, outside and in the book, and there were a lot of them in both places.

It wasn't until I finally figured out the spruce test that I was able to identify that tree by eliminating all the fir, hemlock, yew and redwood pages, not to mention the pines, in the guidebook.

Some say the spruce was named for its tidy, trim appearance. It sounds like a chicken-egg problem to me. Which came first, the tree name or the expression "spruced up," meaning looking well tended? Spruce trees do tend to be shapely trees, usually narrowing at the top. Once you know you've got a spruce on your hands (Ouch!), figuring out which spruce isn't that difficult. There aren't too many spruces in any one area. With three to deal with, New England has more than anybody.

My mystery spruce turned out to be a Norway Spruce, which, although imported from Europe, has been so widely planted that it sometimes appears as common as native species. There's a fairy-tale feeling to this beautiful spruce. It's shaped so differently from other spruces that once I learned about it, I could spot one from a great distance.

THE SPRUCE TEST

Spruce needles are stiff, four-sided and often sharp; fir and hemlock needles are soft, flat and pliable. Spruce needles are the same color on all sides; fir and hemlock needles are lighter underneath.

❧ ❧ ❧

SPOTTING A NORWAY SPRUCE

The branches of a Norway Spruce swoop, either drooping gracefully or turning up at the ends, like welcoming arms. It has the largest cones of any spruce—papery, and up to seven inches long. A fast-growing tree, it can become enormous: Although Rockefeller Center has used a few White Spruces and Balsam Firs, most of the huge trees erected there and festooned with twenty thousand lights have been Norway Spruces.

FEBRUARY 18

Blue Spruce

There's an enormous Blue Spruce behind the local grocery. Its blue color, plus the fact that it's my mother's favorite Christmas tree, makes it the easiest spruce for me to recognize. The tree was so dense that when I circled it, the birds inside simply twitted. They were right to feel safe in their spiky hideout: Blue Spruce needles, up to one and a half inches long, are *sharp*! The top of the tree was thick with stiff, papery cones (three and a half inches), and it passed the spruce test (see page 43).

The Blue Spruce is also known as the Colorado Spruce. Native only to the Rocky Mountains, it is so beautiful that it's often planted as an ornamental tree in many other parts of the country, or raised on Christmas tree farms. I often see it here. In its native environment, it's even easier to spot, for surprisingly few spruces grow there.

❧ ❧ ❧

ONLY TWO ROCKY MOUNTAIN SPRUCES

If you see a spruce in the Rockies—a conifer with bristling, four-sided needles—it's likely to be either a Blue Spruce or an Englemann Spruce, the only other Rocky Mountain native. To tell them apart, look at the new growth on the branch ends: Blue Spruce growth is blue-gray, often quite bright, turning gray green with age, and the needles bristle in all directions; the Englemann Spruce's green needles tend to crowd on the upper part of the branch. The Blue Spruce's papery cones (3½ inches) resemble Englemann Spruce cones, but are about twice as long.

FEBRUARY 20

Black Spruce

I found a young evergreen next to the road a short walk from the house with needles about a half-inch long passing the spruce test. I'm guessing that it's a Black Spruce. I identified this one partly by elimination: There are only nine species of spruce native to the United States, with a few imports thrown in, and only four are native to Michigan. Since the tree wasn't a Blue or a Norway Spruce, it had to be either a Black or a White Spruce. I haven't found a White Spruce yet, but it reportedly has inch-long needles that smell foul when you crush them. Imagine that. I thought all needled trees smelled like Christmas! The needles of this spruce smelled like menthol.

The Black Spruce tends to be a rather tall, thin, scraggly tree with short branches, unpopular for Christmas because the needles fall off soon after it's cut. It likes boggy places best. Its enormous, very northern territory includes most of Canada, dipping south about halfway down Lake Michigan, all the way east through New England.

❧ ❧ ❧

THE SHORT-NEEDLED SPRUCE

The Black Spruce is the only spruce in most of its range—except for New England, where the Red Spruce is found—with really short (½-inch) needles. The half- to one-inch woody cones are another clue: The top edge of the scales are toothy, while the scales on Red or White Spruce cones are smooth.

FEBRUARY 21

Raccoon

A scratchy noise on the windowpane at three this morning sent my heart pounding until I saw the busy-tailed, dog-sized silhouette. Without leaving my bed, I flashed my torch in the un-flinching face of a large raccoon, poised quite casually on the narrow window ledge fifteen feet off the ground. How did it get there? I walked to the window and tapped its nose through the glass. It stared back in, eyes bright and insolent in its black mask.

A thump on the deck took me into the living room. Through the window, my beam lit up another raccoon, then another. Three big, uninvited guests prowled the deck. Not wanting my bird feeders disturbed, I opened the door and stomped out there yelling like a banshee. The startled bandits scampered nimbly along the window ledges, then plunged headfirst down the corner of the house, where cement blocks interlock like log ends. This morning I awoke to find pawprints on the storm door. The rascals had sneaked back. They stole both balls of suet and left open and empty in the driveway the metal cage that held peanut-butter cakes. Although they lack opposable thumbs, those sensitive hands are a challenge. It looks like I have another battle on mine!

THE SEVEN SLEEPERS

Although raccoons are listed among the "Seven Sleepers"—mammals that hiber-nate for much of the winter—raccoons do not really hibernate. They do sleep for weeks at a time, but rise on warmer nights to forage for food. In much of the North, raccoons, chipmunks and skunks are up by the end of February. The other four—bears, bats, Woodchucks and Jumping Mice—usually snooze into March.

❧ ❧ ❧

THREE RING-TAILED MAMMALS

Although no other animal looks quite like a raccoon, raccoons around the country are not all the same. They can vary in size from the three-to-six-pound raccoons found in the Florida Keys to the fifteen-to-thirty-pound variety found farther north. The only similar mammals live in the Southwest: The Ringtail, with a longer, ringed tail, is much smaller, weighing only about two pounds; the Coati, a fairly rare animal, carries its very long, indistinctly ringed tail vertically.

FEBRUARY 23

Dark-eyed Junco

I've been noticing that some of my feeder birds make a separate trip for each seed, flitting back to a branch with it, while others hunker at the feeder and chow down. The chickadee is one of the flitters, and I had often wondered how it came out ahead. It seemed to me that the tiny kernel it finally gulped after flying in, deciding on a seed, bobbling back to a branch and bashing it open could barely make up for the energy spent getting it. The nuthatches and titmice do the same. Cardinals and finches, however, cruise in and feed for long periods, dumping empty shells on the deck.

I puzzled about this to a friend, who explained that it was a matter of beak strength. Cardinals and finches, he said, have big, strong seed-cracking beaks, but the chickadees, nuthatches and titmice have weaker beaks and must pound on the shells to break them. They need an anvil. A kindhearted man, he had also worried about the inefficiency of some birds and had hoped to solve the problem by providing a block of wood inside the feeder, so the birds could break their seeds on that and save themselves a trip to the branch. They ignored his gift and continued launching a separate trip for each seed.

Some birds won't come to the feeder at all; they prefer to peck seeds off the ground—or off the top of the wall. These are usually the bigger birds, like Mourning Doves, Blue Jays and crows, but there's one little bird that will only scratch for food—the Dark-eyed Junco.* It's a distinctive cutie: About finch-sized and slate gray, with a white belly, it stays close to the ground, and when it flies, its tail fans out, flashing white along each side. There are a number of "races" of juncos, and when their ranges overlap, they interbreed. It's still easy to tell a junco, though, even if not quite which junco.

$$\approx \quad \approx \quad \approx$$

ONE EASTERN JUNCO "RACE"

A small, dark bird (5 to 6½ inches) with a sharp demarcation between dark head/body and white breast, making it look almost hooded, is probably a Dark-eyed Junco. When the junco flies, look for white on both sides of the fantail. (Some sparrows have similar tails, but they are usually brown on top, not a solid black or dark gray.) In the eastern states, only one race of junco is present, and then only in winter. In the West, several types may interbreed.

FEBRUARY 27

Canada Goose

About ten days ago, for the first time this winter, it was no longer dark when a friend and I drove to our early morning lap swim in Holland, about twenty minutes north. The light was delight enough, but strewn across the brightening sky were skeins and skeins of geese. Some were distant, dotted Vs, but others flapped right over the highway, honking loudly. We became very excited. It looked as if the Canada Geese were beginning to migrate! Surely this was an early sign of spring.

Today I visited a nearby refuge for geese. The drive took me about fifteen miles southeast of Douglas, along quiet country roads flanked with

---

* Sparrows like ground-pecking, too, but I never got many of those.

farms, before I came to the headquarters for the Allegan State Game Area, a barn and a small white building sitting next to a creek between dry, shorn cornfields. The building looked as if it were locked, but a man in a DNR* cap heading for the road on a large green tractor suggested I bang on the door. This was answered by a thin, white-haired man who informed me gently, as if to cushion my disappointment that spring was not yet here, that the geese did not begin migrating until later in March. For more than sixty years (thirty of them as an official, thirteen-hundred-acre refuge), the farm had been a spring and fall stopover for thousands of migrating Canada Geese—sometimes over twenty-five thousand occupied the place at one time—but a smaller flock of Giant Canada Geese also wintered there. What I had seen that morning was not a spring migration, but the daily dawn flight of wintering geese from their nighttime quarters on surrounding wetlands to their daytime feeding grounds on the farm, attracted by long tradition and the portion of corn, rye and other grains left by sharecropping farmers.

"But where are they?" I asked. I had not seen one goose in all the acres of mown cornfields on either side of the road.

"Oh, they're around," assured the man. "They're around."

I walked up the road to find them. A few crows flew by, cawing. A treeful of blackbirds screamed. No geese. The tractor I'd seen as I'd arrived was heading away from me across a large field slightly uproad. Suddenly thousands and thousands of geese flew up, honking furiously, from the far end of the cornfield, darkening the sky like swarming bees. Arcing high and swirling overhead as if mixed by a large spoon, they dispersed in several directions after about ten minutes of dizzying indecision. The air felt suddenly chilly. Except for the blackbirds, the place seemed again like any other farm waiting out the winter.

Migrating Canada Geese, usually welcome heralds of spring, can become a nuisance when they tamely strut their stuff in parks and on beaches. One desperate researcher determined that a Canada Goose eliminates every four minutes! They can be bold, too: A friend told me yesterday he'd seen a flock of them enjoying a small pond at a General Motors research facility in industrial Detroit!

* Department of Natural Resources

୬ ୬ ୬

ELEVEN "RACES" OF CANADA GEESE

Of course you know a Canada Goose—big (16 to 25 inches), long-necked, black feet, neck and face, with that distinctive white patch, like a round-ended bandage, that reaches from under the chin more than halfway up the head. They are seen all over North America. But did you know there are *eleven* "races" (subspecies) of Canada Geese? Varying mostly in size, the smallest (2 to 3 pounds) looks like a Canada Goose, but has a stubby neck. The largest (13 to 18 pounds), the Giant Canada Goose, can grow as large as an average raccoon. Other races place in between. Lacking better terms, even scientists refer to them as small, medium, and large.

FEBRUARY 28

Red-winged Blackbird

Maybe the geese weren't doing a spring thing yesterday, but I have it on good authority that the hundreds of Red-winged Blackbirds yodeling their hearts out in the top of the tall, bare-branched tree nearby certainly were! I know a Red-winged Blackbird when I see one—who hasn't seen them swaying on cattails and singing their throaty swampsong?—and this tree was practically in leaf with them. They were all black, robin-sized birds preening their crimson and gold shoulder epaulets and announcing spring in a most unharmonious chorus. The familiar *O-ka-LEEEEEE* could only be heard during occasional pauses.

But something was very strange here. I hadn't realized that the female isn't black at all—she is brown, with a heavily streaked breast, rather like a large, female House Finch—and I could see no brown birds in that tree. When I looked up Red-winged Blackbirds in my bird book, I learned that the birds I was observing were all males!

Further research disclosed that I had not found a fluke. The male Red-winged Blackbird, not the robin, is often the first bird of spring. Late in

February (in much of the North), male Red-winged Blackbirds fly in large flocks back to their nesting grounds to spend two or three weeks staking out their closely bordering territories, before the females fly up to choose their mates. The Red-winged Blackbird male, unlike most songbirds, often takes several females into its territory. With so many mates and young to protect, he can become quite fierce. A friend who cycles vigorously says she's frequently dive-bombed by Red-winged Blackbirds when she wheels along the wetlands.

ᴓ ᴓ ᴓ

ONLY ONE RED-WINGED BLACKBIRD LOOK-ALIKE

The red and gold (or buff) shoulder epaulets give away the male Red-winged Blackbird, as well as its loud, liquid song: *O-ka-LEEEE!* or *Conk-a-LEEEE!* Only in the West Coast states is there a confusing species: the Tricolored Blackbird, which also has red epaulets but with a white lining. It's easier to distinguish by its song, however, which is quite unmusical.

March

Bird Songs

 It is March. I usually wake up to March with dread, sick of rattling branches and metallic skies. It never feels like spring to me until late April, when a faint green haze softens the landscape. This year, however, I am delighted by the absence of leaves, because early March is a wonderful time to learn bird songs.

On warmer days, the wintering birds have begun to sing. A cardinal awakened me at dawn, chortling *Birdy birdy birdy birdy* outside my window. When I stepped outside, I heard scattered announcements in the trees. If I listen carefully, I don't hear just tweets and twitters, as I always have before. I hear several quite distinct, quite different songs. Bird song in early March is chamber music. The masses of transient migrators and the summer nesters have not yet appeared, so I am not barraged by a full orchestra. Birds have few places to hide, so I can begin to match the song to the singer.

It has taken much careful listening. I just think I know a song, and the next day, I can't remember it. Slowly, however, I begin to sort out the music. It's wonderful to walk down the road and be able to see, in my mind's eye, what I am hearing, even when the singers are out of sight.

🐦 🐦 🐦

EARLY MARCH BIRD SONGS

Words don't begin to communicate bird songs, but without electronic help, they will have to suffice. Further complicating matters is birds' propensity for variety: You just get a species song down and the bird sings something else. Nevertheless, it is possible to pick out some of the simpler calls. It's easiest to do this early, before the robin arrives with its virtuoso's repertoire of spring arias.

SONG	DESCRIPTION	BIRD
Chick-a-dee! *Chick-a-dee-dee-dee!* *Dee-dee-dee!*	Double-noted, hoarse	Chickadee

SONG	DESCRIPTION	BIRD
Nyant, nyant!	Hornlike tooting	White-breasted Nuthatch
Screeee! Screeee!	Loud, hoarse scream	Blue Jay
Birdy birdy birdy birdy birdy!	Loud, rich call	Cardinal
Chip! Chip! Chip!	High and clipped	
Peter, Peter, Peter!	Loud whistle, accent	Tufted Titmouse
Pete! Pete!	on either syllable	
Caw! Caw! Caw!	Harsh, falling call	Common Crow
Coo! Coo!	Soft, falling notes	Mourning Dove

MARCH 3

Muskrat

The robin is fast losing its distinction for me as the first sign of spring. Today, at a lagoon formed by a dead-end branch of the river just short of the lake, I saw a muskrat on the ice, devouring what looked like a catfish. I came within a stone's throw of the furry, football-sized creature before it slipped into a nearby hole, taking its scaly tail and its hoagie-sized meal with it. Seconds later, it returned, meal intact, with company. I was not alone, watching the muskrat pair. Several crows cawed hungrily from nearby trees on the other side of the lagoon, and a pair of greedy gulls swooped in again and again, but the muskrats were not to be distracted. They sat tight and ate, even when I got up and walked farther up the shore.

Muskrats bringing food up through "push-up holes" in the ice, I read later, is an early sign of spring. Although they don't hibernate, muskrats tend to winter in their lodges, many of them sometimes crowding together for warmth, and feeding mostly under the ice (they can remain underwater for up to fifteen minutes!). I had seen several of their domed, thatched mounds on the other side of the lagoon, and thought they were beaver lodges, but muskrats make them in fall from some of their favorite foods:

cattails and grasses. They excavate rooms from an underwater entrance and, in late winter, devour the walls if the food becomes scarce.

What a rich and varied natural area I inhabit! Within a few minutes' walk or drive, I can explore dunes, forests, wetlands, the Kalamazoo River, ponds, creeks, or Lake Michigan—each with its own look and life.

∂∂ ∂∂ ∂∂

MUSKRAT TAIL

While a beaver's tail is flat as a pancake, a muskrat's tail is vertically flat, like a side-winding snake. Although there are other small furry animals with naked, ratlike tails, none are flattened this way. (The smaller Florida Water Rat and the larger Nutria have round tails.) If you see a V-wake in the water of ponds or quiet river backwaters, look for the tail. If it's a muskrat, you can see it; you can't see a swimming beaver's tail.

MARCH 4

American Goldfinch

My new finch feeder has become a main attraction on the deck. The squirrels leave it alone, the raccoons have not yet managed to steal it, and finches have been perching on it several at a time, red ones with their brownish companions and some small, olive-colored birds. This afternoon, I exclaimed over some similar but patchy yellow finches at my neighbor's feeder. She observed, "You can tell it's almost spring—the goldfinches are starting to get their color back!"

That explained everything. The small, dull-colored bird with the yellowish head was beginning to molt, turning a brilliant canary-yellow, with the exception of its black cap, wings and tail. I saw a whole flock of these goldfinches several weeks ago along the river and thought they must be some sort of migrating birds, but goldfinches like to hang out in big flocks,

---

A GOLDFINCH SONG

In addition to their canarylike warbling, goldfinches often sing *perchicoree!* during bursts of flight.

---

warbling their high, sweet music, looking and sounding much like wild canaries. These flocks cruise the countryside all over the United States in winter, and much of the North and Canada in summer.

Although they do eat insects part of the year, goldfinches love seeds, especially seeds attached to puffs, like Goat's Beard, dandelion and thistle seeds. Late maters, the goldfinches wait for these plants to seed out in late or midsummer before nesting; they pad their nests with the down.

GOLDFINCH CLUE

No other small (4 to 5 inches) yellow North American bird has a black cap, wings and tail. (The Evening Grosbeak resembles its coloring, but is much bigger.) The duller-colored female isn't speckled, like female House and Purple Finches. Males molt in fall, becoming an olive color, but retaining the black-and-white wings.

MARCH 5

Red-Osier Dogwood

I was with an artist friend, driving some paintings to a show, when I noticed clumps of a shoulder- to head-high, beet red–branched shrub along soggier sections of roadside. I'd been wonder- ing all winter what these were. Often I watch the sun rise through their smooth, slender, bril- liant branches, which rise from the boggy lot across the street, and which have appeared to me lately to be brightening to alizarin, a painter's purplish red. Having no shrub guides and intimidated by shrubs anyway, especially

VERSATILE DOGWOOD

A dogwood can be a tree, shrub or wildflower. The Flowering Dogwood is a tree; the many dogwood shrubs include the Red-Osier Dogwood; and the Bunchberry is a low-growing dogwood wildflower. All dogwood leaves are *opposite** (leaves that grow in pairs, across from one another), with prominent leaf veins that curve upward in pairs, like an elaborate, ancient candelabra. Large Dogwood flowers have four *bracts*, which look like petals, but are actually modified leaves. (The bracts on most flowers are small and green.) The tree blossoms are larger than the wildflowers, and the shrubs blossom in tiny flowered clusters.

* The one exception is called the Alternate-leaved Dogwood.

leafless shrubs, I hadn't attempted to identify them. Ellen, however, hearing me exclaim over the bright branches, required neither leaf nor second thought. "Oh, that's Red-Osier Dogwood," she remarked. "It's everywhere!"

I thought a dogwood was a prized tree, not a common bush—but I was wrong.

❧ ❧ ❧

SPOTTING RED-OSIER DOGWOOD

Look for red, eight- to twelve-foot, smooth, slender branches in damp areas. The intensity of red depends on sunlight: the more sun, the brighter the branch. (The other red branches I've found here in winter are thorny brambles of a darker, duller red.) Other Red-Osier clues: opposite, two-inch, flame-shaped leaves; white flower clusters (spring); white berries (fall). It is found all over the northern half of the country and Canada.

MARCH 8

"Snags"

Today it rained, a cold, nasty rain, but I took a short walk anyway, under my new hot pink umbrella, looking for mushrooms. On awful March days, I enjoy reading about what the weather prevents looking for, and during the last several days of relentless rain, I had learned that a good time to look for mushrooms is after several days of steady precipitation. I didn't find any mushrooms, but I did see some white foam, which I supposed to be maple sugar sap dripping from woodpecker holes, frothing at the foot of quite a few tall trees. I also came upon three freshly cut tree stumps standing next to the road like large, rejected coffee tables, wet tops slick and glossy.

I'd also been reading about dead trees, called "snags," and the sight of the stumps saddened me terribly. A one-hundred-year-old tree can take another hundred years to disintegrate completely, attracting insects which attract insect-eating birds, and softening so that cavity-nesting birds—about 10 percent of all species of North American birds—can excavate homes. Many species of birds, especially woodpeckers, cannot survive without such trees. Dead trees, once culled without thought, are now officially appreciated: The U.S. Forest Service and many state forest services have adopted policies requiring timber companies to leave a certain number of such trees standing to support wildlife. No doubt, my neighbor's trees were removed as hazards to traffic along the road, but I've begun looking for dead trees and finding them full of life.

❧ ❧ ❧

DEAD TREES SUPPORT LIFE

Many naturalists are urging homeowners to leave as many dead trees as are safe (not a threat to road or home) rather than discard them or use them for firewood. Snags attract many kinds of birds and animals, providing food and homes for some species of ducks, owls, songbirds, woodpeckers, salamanders, even small and medium-sized mammals.

MARCH 9

Pine Grosbeak

For weeks, a shelf of ice has reached so far out on the lake that, standing at the edge, I cannot hear the waves breaking. This is not the lidlike skating ice that often caps a smaller lake. Lake Michigan ice, at least on our eastern shore, is a hilly, cavernous, treacherous snowfield that with wind and weather changes shape daily, from a cratered moonscape to a peaky, meringue-topped pie to what looks like the aftermath of snowball-cannon warfare. I walk it often but rarely see any life there—plant, animal or, for that matter, human.

Today the ice began breaking into white boat-sized bergs bobbing in fiercely blue waves, and I could walk on sand again. The temperature rose into the low forties and the sunlight fractured off the floating ice as I looked for something I could name. Alas, I saw only Herring Gulls, so I climbed a friend's steep beach stairs and began walking down a country road toward town, past quiet fields and orchards on one side and some scrubby woods on the other.

It felt good to be outdoors. Cardinals and titmice sang as if in competition, flashing in front of me every few minutes. A pool-sized puddle in a stubby cornfield held a pair of Mallard Ducks. A Red-winged Blackbird yodeled aggressively from the tip of a spiky tree. I'd almost reached the Blue Star Highway, which sweeps through the middle of Douglas, when I heard a great deal of warbling and whistling. Three huge spruces, taller than the two-story house they flanked, were aflutter with red birds and yellow birds. They looked like House Finches and American Goldfinches having a party. Did two species hang out together like this?

When I got closer, I realized that the birds were nearly twice the size of small finches. They ignored me as I stood paging through my bird book, a tolerance I appreciated, for it took me quite a while to sort through all the red and yellow finches. I found three species of finches that had red males and yellow females, but finally the large size left me with only one possibility: Pine Grosbeaks.

ᔥ ᔥ ᔥ

PINE GROSBEAK: THE BIGGEST FINCH

A robin-sized (8- to 10-inch), red-bodied bird with no crest and a yellow mate is likely a Pine Grosbeak. A northern species, the Pine Grosbeak, a name which means "stubby-beaked," is the biggest finch in the eastern United States. The male resembles a large House Finch, while the female is olive-colored or a dull yellow. Two species of crossbill also have red males and yellow females, but they are only slightly larger than the House Finch, and the ends of their bills cross, like scissor blades.

MARCH 10

Staghorn Sumac

My friend Marcia, who had taken a similar walk yesterday, proposed that our outings had not been spring walks, but *looking-for*-spring walks. I told her I'd seen a robin. "Robin or no robin, it just doesn't *feel* like spring yet," she said wearily. Some of the things I take to be signs of spring—like flocks of Pine Grosbeaks or Canada Geese—are not spring signs at all, but winter phenomena I don't know enough about.

I expect that the fuzziness I saw on the sumacs along the road, which I'd hoped was another sign of spring, fit into that category. As I walked home yesterday, my eye was caught by the beautiful skyward curves of leafless sumac branches. They were antlerlike, deliberately drawn, like the smiling, scalloped lines of an early Mondrian painting. I knew these were sumacs from the dry red fruits—each a pomegranatelike collection of fuzzy seeds, some almost as big as my hand—flaming from the branch ends, but I didn't know which sumac. I resisted the temptation to stroke the velvety branches, thickly covered with soft hairs. Some sumacs, I knew, are poisonous.

What I'd seen was, however, a Staghorn Sumac, named for its curved branching and its velvet bark. The poisonous variety was easier to tell than I realized.

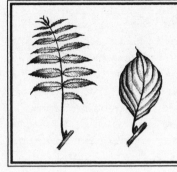

HOW TO TELL A COMPOUND LEAF

Every leaf on a tree or shrub has a little leaf bud at the bottom of the leaf stalk. A single leaf that has this little bud is a *simple leaf*, but a "leaf" lacking this bud is actually a *leaflet*, part of a group of leaflets that make up a *compound leaf.* Sumacs have compound leaves.

🙚 🙚 🙚

POISON SUMAC VS. STAGHORN SUMAC

A shrub with fuzzy branch ends and clusters of fuzzy red fruit is not Poison Sumac. There are at least seven common sumac shrubs, all with compound leaves and oval leaflets, and some with fuzzy red fruit. Staghorn Sumac, however, is the only sumac with fuzz on new branch growth. Poison Sumac is the only sumac with white berries. Its leaflets have smooth edges, while most sumac leaflets are saw-toothed. Do not touch any part of Poison Sumac! (See page 148.)

MARCH 12

Eastern Chipmunk

On this aquamarine day, I went out on the deck to watch the icebergs on the lake splitting into bits that float there like fallen fluff, and nearly stepped on a chipmunk no bigger than my hand, from nose to its chomped-off tail. It sure was cute, sitting a foot from my foot, looking hopefully up at me, cheeks bulging. When it suddenly panicked, it couldn't seem to climb the three-and-a-half-foot wall that enclosed the deck like half of a large hot tub.

HOW TO TELL A CHIPMUNK FROM A SQUIRREL

Most people look at the back stripes, but look for white face stripes first. You can't always tell a chipmunk from the stripes on the back, for some squirrels also are striped. No squirrel, however, has the chipmunk's white stripes over and under the eye. This is how Easterners can tell an Eastern Chipmunk from the Thirteen-lined Squirrel and Westerners can tell the Least (or other) Chipmunk from the Golden-mantled Squirrel: Neither of the squirrels has white eye streaks.

I found a big grocery bag and tried to shoo the chippy into it so I could release it on top of the wall. The chipmunk got so scared that it lost its precious bounty, dribbling a six-inch trail of seeds, but would not run into the bag. Finally, I propped up an eight-foot branch of driftwood. Zip! Quicker than my eye could follow, the chipmunk had flown off the wall and was gone, fifteen feet below.

After much deliberation among the surprising number of look-alike chipmunks presented in my mammal guide—there are about twenty species of chipmunks in North America, plus a few chipmunklike squirrels—I discovered that my guest was, without doubt, an Eastern Chipmunk. If I lived in the western United States, I might still be playing "What's different about this picture?" In most of the eastern United States, however, there is only one common chipmunk—the Eastern Chipmunk. Its only serious competition is the Thirteen-lined Squirrel, which, like several squirrels, looks like a chipmunk but isn't.

The Eastern Chipmunk is one of the early rising "Seven Sleepers" (see page 46) but, like the raccoon, wakes up from time to time during winter. By this time in March, however, the chipmunk is up for good.

ᏒᎡ ᏒᎡ ᏒᎡ

EASTERN OR LEAST CHIPMUNK?

The Least Chipmunk looks like a smaller version of the Eastern Chipmunk. The two are hard to tell apart, but their ranges overlap only around the states and Canadian provinces surrounding Lake Superior. If you see a chipmunk there, look at the base of the tail: The Least Chipmunk's stripes reach the tail; the Eastern Chipmunk's don't.

Eastern Diamondback Rattlesnake

Although I did see this snake in the wild, I did not find it in my neck of the woods, thank heavens. A few days ago, I drove to the coast of Georgia to vacation with my parents, my college-age son, who flew in from Wyoming, and my British cousin Richard, a "solicitor" on his way home from a Bird Questers' tour in Chile. We were an odd lot, to be sure. Eager to please everybody, including our birder guest, my father arranged an outing to a small island which we reached by outboard motorboat. We saw many birds on the way, including pelicans, gulls, fishing eagles and some comical, gourd-bellied birds perched defiantly on No Hunting signs. Once there, we were trucked, with other tourists, around the island, over bumpy dirt roads, our dozen bottoms thumping on boards set, unattached, over the beds of each of two pickup trucks.

During our frequent stops, I peered at the literally thousands of shore-birds dotting the lovely beaches, through Richard's telescope, which he carried with him always, along with a tripod and a case heavy with binoculars and camera equipment. We saw strange animals, too. Large, motionless, mud-colored alligators dozed on the creek banks; tiny, white deer appeared, dreamlike, for brief instants between the trees and low, fanlike palms; and several muscular rattlesnakes writhed from forked sticks held by three men in khaki shorts, knee socks, short-sleeved shirts and hats, who were, they told me, from the police force. I first saw these men on the dock when we arrived, and hearing that they were planning to hunt rattlesnakes, I went over and inquired why. A guarded look leapt into their eyes. The snakes were for medical laboratories, they said.

Later we came across the men at the other end of the island. At the request of our guides, the officers took two snakes out of a big white bucket, handling the poisonous creatures deftly with a forked prod and their hands. The diamondback bodies were as thick and long as a lanky man's leg. Later, our guide told us that the snakes were being caught for a Rattlesnake Roundup. These roundups, a popular southern spring rite featuring "Big-

---

NO MORE OPHIDIOPHOBIA!

You're probably more likely to drown in your bathtub than you are to die from the bite of a poisonous snake. Despite most Americans' ophidiophobia—fear of snakes—according to the *Encyclopedia Americana*, fewer than ten people a year die from snakebite in this country, and probably half of those are snake handlers. In contrast, it has been estimated that thirty to forty thousand people in Asia die annually from snakebite.

---

gest Rattlesnake" and other contests involving the hunting, comparing, deep-frying, eating and killing of hundreds and hundreds of specimens, are at last being questioned by environmentalists. This changing public attitude explained the suspicion I inspired at the dock.

At first I thought I'd identified the snakes—rattlesnakes, right?—until I discovered that there was more than one kind of rattlesnake to choose from.

By a process of elimination, I was able to identify the island rattlers as Eastern Diamondback rattlesnakes, the largest of all the rattlers. Their size and diamond-pattern design were, unfortunately for them, a dead giveaway.

🐌 🐌 🐌

THIRTY SPECIES OF RATTLESNAKES

About thirty species of rattlesnakes, plus many subspecies, range the Americas, from Canada to Argentina. There are about sixty names in all. Here is a rough distribution of North American rattlesnakes (some have small ranges):

NORTHWEST	*WESTERN GREAT LAKES*	*NORTHEAST*
Prairie	Massasauga	Massasauga
Northern Pacific		Timber

WEST COAST	*GREAT BASIN*	*ROCKY MOUNTAINS*
Southern Pacific	Midget Faded	Prairie
Northern Pacific	Great Basin	
Red Diamond*		
Speckled*		

SOUTHWEST

Massasauga	Arizona Black	
Twin-spotted	Speckled*	
Black-tailed	Mohave	
Tiger*	Mohave Desert	
Rock*	Sidewinder	
Ridge-nosed*	Sonoron Desert	
Prairie	Sidewinder	
Grand Canyon	Colorado Desert	
Western	Sidewinder	
Diamondback		
Hopi		

SOUTHEAST

Eastern
  Diamondback
Western Pigmy
Eastern Pigmy
Dusky Pigmy
Canebrake

* Range is mainly in northern Mexico.

HOW TO TELL A POISONOUS SNAKE

An easy-to-see sign of any North American poisonous snake (except the Coral Snake,* found in a small part of the Southwest) is a head that is wider than the neck, especially a heart-shaped head. Poisonous snakes in the United States, except the Coral Snake, are *pit vipers*, with wide heads. The other signs are harder to see: Pit vipers appear to have a second pair of nostrils, used to locate prey, placed forward of and under the eye. They also have vertically elliptical irises. If you are close enough to see the pits of its nose and the slits of its eyes, back off!

* There is a warning rhyme about the striped Coral Snake: "Red touch yellow: Poisonous fellow!" Some Coral Snakes can be all black, however.

MARCH 16

Wood Stork

I am convinced that even rank amateurs like me will be rewarded with a glimpse of something rare and wonderful if we poke around nature in a regular sort of way. My reward was a group of Wood Storks, found on our island tour. These big-billed, bald-headed, white birds, with black-bordered wings and tails, hunkered together a little like vultures with a flock of white egrets in a rookery. Early in this century, more than a hundred thousand Wood Storks nested in groups numbering in the hundreds, even thousands, in over thirty swamps in Florida, but cypress logging and development has destroyed most of their nesting grounds. Today, our only North American stork has the sad distinction of making the federal government's endangered species list.

There were about a dozen Wood Storks in all. I never could have distinguished the storks from their numerous egret neighbors, however, if Richard hadn't lent me his binoculars. Through these, I could make out the black, featherless heads, the long, strong bills and the black-edged wings.

Richard is the first really serious birder I have ever seen up close. I tend to think of birders as a bit dotty: obsessive persons who subject themselves to discomforts of all kinds, such as rising before dawn or carrying a great deal of equipment on an otherwise pleasant hike. Richard did fit my caricature, but he had a sense of humor about it. He insisted that his symptoms were positively mild compared to many birders in England, where birdwatching often takes on a manic quality.

"There are birders in England who live for the sole purpose of ticking off another bird on their life lists," said Richard, dropping his *r*s and crisping his consonants. He explained that sometimes migrating American birds are blown off course and over the coast of England by a storm. When one of these birds is sighted, the exciting news goes right onto the radio. Devoted birders will travel a hundred miles—considered a huge distance in England—to catch sight of a new bird and tick it off on their lists. This chase is known as "twitching." If birders go all that way and don't see what they went for, they have "dipped." "It is a terrible thing for a twitcher to dip," concluded Richard mournfully.

🐦 🐦 🐦

ENDANGERED BIRDS: WHO'S WHO

Passed in 1966, the Endangered Species Act protects species in jeopardy. By 1986, thirteen species of birds were listed (as well as eleven subspecies):

Wood Stork	Peregrine Falcon
Whooping Crane	Red-cockaded Woodpecker
Brown Pelican	Ivory-billed Woodpecker
Eskimo Curlew	Bachman's Warbler
Least Tern	Kirkland's Warbler
California Condor	Piping Plover
Bald Eagle	

MARCH 18

Elephants, Camels
and Lions, etc.

This is not a trip to the zoo. Yesterday my father took us to the University of Georgia Marine Extension Service, a seaside lab where Taylor Schoettle told nonstop stories about objects I first took to be stones crowding two large glass cases outside his pleasantly cluttered, green-tiled office. The "stones" turned out to be bones. "There are no rocks natural to the Georgia coast," said Taylor. "For a hundred miles, nearly halfway into Georgia, you will find only sand, bones and shells. The continental shelf is drifting toward the west. That's why the West Coast has cliffs and rocks, but here on the dragging side there are no rocks at all."

Out on the dock, we poked through a bin full of black bone fragments, some as large as forearms, dredged up by students using two confiscated drug-runner shrimp boats. I picked up some pieces and found they had a surprising and satisfying heft, like paperweights. Taylor gave me two that clinked musically against each other. The one probably came from a whale rib, he said, while the other might have been part of a sea turtle carapace. The satiny black, petrified bones, one in each of my hands, felt cool and mysterious. I sensed in them the deep, ancient sweep of the sea.

Taylor led us back to the cases, continuing with easy, happy enthusiasm about their contents. "Less than thirty-five thousand years ago, when humans were probably already present, there were still enormous animals living here." There was a three-hundred-pound beaver, he said, and a twenty-foot sloth that "ate trees like carrots." There were also several kinds of elephants, a mammoth, and even a lion that resembled the African lion, roaming the Georgia coast.

Moving on to the two shelves of shark's teeth—one tooth was as big as my hand—Taylor explained that sharks don't have teeth. Shark's teeth aren't teeth at all in the usual sense, but are really modified scales. He handed me a concave patch of shark's skin, which felt like an emery board. Shark teeth, Taylor explained, grow in several rows and fall out easily when a shark bites into something, its teeth falling through the water "like snow-flakes." He smiled, his hand drifting gently before him.

I asked Taylor why the water along the coast often looked like cola, dark and reddish, frothing behind motorboats. "Two things are involved there," he replied. "First, a red algae, as well as tanins from leaves and bark, tend to make the water look red, but the real reason the water is so murky is that it is teeming with life."

MARCH 20

Binoculars

On my second evening back home, I experienced an urgent, immediate need to buy a pair of binoculars. I wanted to know *tomorrow* what sang that pretty dawn song outside my bedroom window. I soon found myself entering one of those enormous open-till-nine everything stores, and I did what amateurs often do: analyzed limited options with incomplete information, jumped in and hoped for the best.

I probably was working with more information than most, however: When I'd been in Georgia, I had borrowed as many binoculars as I could, noting that I tended to see more clearly through the bigger, heavier models. As well, Richard had advised me not to buy a cheap pair, but to spend at least a hundred dollars. Finally, I had written a short piece about binoculars in my last book, *Reading the Numbers,* in which I struggled to make sense of measurements and sizes, so I knew what the size numbers meant, which, as it turned out, put me way ahead of the salesperson.

The young man who waited on me had never looked through a pair of binoculars in his life, but he was polite and patient while I tested each of the six models of Bushnell binoculars, on people at the other end of the vast aisles. I finally decided on the heavier 10 × 50 model, because the "picture" seemed so much brighter than the lighter 7 × 35. They cost only

---

IF YOU WEAR GLASSES

Binoculars with special rubber fold-backs on the eye pieces are available for people who wear glasses. My binoculars have these, but I find I still see better when I take my glasses off. Don't lay your glasses in the grass: I did this once and nearly stepped on them. Shove your glasses up on your forehead or put them on a chain.

---

eighty dollars, but I'd noticed that the naturalists in Georgia used that brand, so I assumed that they must be acceptable.

READING BINOCULAR NUMBERS

The first number on a binocular size tells the number of times the pair magnifies; the second number is the size of the lens. A pair of 7 × 35 binoculars magnifies 7 times through a 35mm lens. You can buy medium-priced binoculars that magnify 7 to 10 times, depending on the pair. More expensive binoculars can go up to a magnification of 15. Often, however, the greater the magnification, the greater the weight.* Lenses vary from 35mm to 60mm. The larger the lens, the wider your field of vision. Large lenses also help you see in less light (birds are most active at dawn and dusk).

* For a greater price, excellent, lightweight binoculars are available as well.

MARCH 21

Red-breasted Merganser

I awoke to that dawn song that's been driving me crazy. *FEEEEEE-beeeeee . . . FEEEEE-beeeee . . . FEEEEEE-beeeeee*, the high, clear *FEEEEE* about two whole notes higher than the *beeeeee*. I looked up the phoebe in my bird book, but I didn't recognize it. Phoebes are flycatchers, unlikely guests at my feeders, so a little later I went phoebe-hunting with my new binoculars.

HOW TO CATCH A BIRD IN YOUR BINOCULARS

This is trickier than you think: If you aim your binoculars at the treetops, one up-close mass of leaves or branches looks much like another. I have found it easier to locate the base of the occupied tree first, then move the binoculars up the trunk and out the appropriate branch. With luck, the bird is still there.

I walked for about half an hour, but I couldn't find a phoebe any-where. I had a hard time finding anything, actually; my new, large-sized (10 × 50) binoculars hung so heavily around my neck that it took a special effort to look up. Furthermore, using the binoculars was sur-prisingly difficult; I would see a bird in a tree, take off my glasses, look through the binoculars, focus, try to find the right tree, then the bird, fail, and start all over again, which meant putting my glasses back on, since I am very nearsighted. Eventually, I learned to push my glasses up on my forehead, but even then, it was not a graceful, simple pro-cedure. I wasn't particularly bad at this, I learned later. Taking quick, accurate aim with a pair of binoculars can require the skill of a sharp-shooter!

I was just scolding myself for waiting so long to go out—I'd lost the early morning sunshine—when I saw a flock of what I thought were Canada Geese land on the lake. With growing excitement, I realized that the geese I'd seen before always flew in formation, loud and low, never landing near the shore. These birds were silent, and when I finally found them in my binoculars, they had settled on the water very close to shore. Standing above them on a dune, I had plenty of time to alternate focusing through binoculars and flipping through the pages of my bird book. After about five minutes of this, I concluded that I was looking at more than fifty Red-breasted Mergansers.

I'd never seen a merganser, not in all the summers I'd spent on the shores of Lake Michigan. The mergansers were about the same length (twenty-three inches) as the Herring Gulls that floated among them. Some of the mergansers were marked in splendid black-and-white patterns in the impressive way of loons, but showy black feathers sprayed from the backs of their heads like hair in a stiff wind. The females were chestnut brown and extremely beautiful, also with a crested head. Was I seeing these exotic birds with my own eyes in my own surf? Surely I was back in my house watching the Discovery channel!

I sat on a stump in the chilly March air for about half an hour, watching the beautiful mergansers fly in, fly out . . . swim as if for their lives, necks stretched way out in front of them, feet paddling to beat the band . . . dive for fish, disappearing completely, to come up who knew where.

🙚 🙚 🙚

WHAT IS A MERGANSER?

Mergansers are diving, fishing ducks with the hard-to-see special feature of a serrated bill, but they are probably easier to tell from other ducks by looking at the female. The female of all three species of merganser common to North America has an elegant crest at the back of her rusty-red head. All three mergansers are found around the Great Lakes and both East and West Coasts.

MARCH 23

Great Horned Owl

I'm beginning to believe that there is nothing particularly lucky about nature writers and photographers. All a person has to do to see something special in nature is be there. A little knowledge sometimes helps, too, as it did today, when I was padding through a nearby woods and heard a bunch of crows squalling around the top of a big hemlock. I remembered that business about crows mobbing owls in the daytime (see page 24), so I thought this might be a good time to see if it checked out. It took me about fifteen minutes to get into a place where I could actually see the owl, but it was there, all right, and it was a big one. I could even see its "ear" tufts, which aren't ears at all. No doubt about it: The crows were after a Great Horned Owl, probably their worst enemy.

One reason owls keep quiet in the daytime is to avoid attracting the harassment of crows and jays. This owl was not so wise. Not five minutes after I arrived home, my friend Marti stopped by for tea and told me she had seen the same owl in the same place only a week ago. "I stepped into the middle of the clearing there," she said, "and as I did, an owl hooted, loudly and quite near, three times, then three times again, then three times again."

The Great Horned Owl is one of the fiercest hunting birds in North America, picking off crows with ease from their nighttime rookeries and preying on mammals as large as porcupines, rabbits and skunks, other owls, hawks and even Great Blue Herons.

ë&& ë&& ë&&

WHICH BIG, DARK-COLORED OWL?

There are only three really big, dark owls common to North America. The Great Horned Owl (up to 25 inches) is found everywhere; the Great Gray Owl (up to 33 inches) is usually found in the Northwest, near the Canadian border and in Canada; and the Barred Owl (up to 24 inches) lives in the eastern half of the United States. In the West (but not the Northwest), an enormous, brown owl is likely to be a Great Horned Owl. In the East, it's either a Great Horned or a Barred Owl. The Great Horned Owl has "horns," while the Barred Owl doesn't. (See also page 242.)

MARCH 24

Nodding Wild Onion

Looking for something green, something growing, today I found numerous patches of emerald-leaved wild onions along the road—slender, erect shoots that looked lush and edible. Once I recognized them, I began seeing them everywhere, so I guess I must have been blind all these years, never having seen the stuff before. Friends have since told me that wild onions grow more frantically in their yards than dandelions.

Wild onions look much like chives, but their grasslike leaves are thinner, more delicate. Picking some, I could smell that distinctive onion aroma. I couldn't tell without flowers exactly which kind of wild onion I had found, but after reading my wildflower books, I guessed—from the slightly flattened, hollow leaves—that it was Nodding Onion, the most common variety found here, named for its cylindrical, purple flower heads that nod instead of poising themselves like knobs atop their stalks.

Some sort of wild onion or garlic is found in most of the country. According to one source, *Chicago* derives from the Ojibwe word *She-ka-kong*, Place of the Wild Onion. Another source said it meant "Smelly place," referring to Wild Onion. Deer and cattle are especially fond of wild onions, which can make their milk taste sour. Bears with garlic breath are not unknown. Human beings snip the spring greens into salads, boil the bulbs or use them for seasoning.

ﾞﾞ　ﾞﾞ　ﾞﾞ

SNIFFING OUT THE WILD ONIONS

If a plant looks like an onion (grasslike leaves, white flower cluster on top of a stem) but doesn't *smell* like onion, it could be the poisonous Star-of-Bethlehem or Fly-poison. Fly Poison is so lethal that failing to wash before eating after handling the plant has reportedly killed some unfortunate people. Wild onion, garlic or leek smell strongly of onion or garlic, and all are edible. Nearly all have pink or white flowers, but you can tell a garlic by the bulblets mixed with the blossoms.

SPECIES	WHEN IT FLOWERS	FLOWER COLOR
Nodding Wild Onion	July to August	Pink or white
Wild Onion	July to August	Pink
Prairie Onion	July to September	Pink to lavender
Field Garlic	May to July	Pink or white, with bulblets
Wild Garlic	April to July	Pink or white, with bulblets

MARCH 26

Brown-headed Cowbird

There's an arsenal of rolled-up socks piled by my desk today, which I hurl from time to time at the annoying cowbird that has been raiding my feeders on the other side of the window. It's not an unsatisfying activity: The windup, aim and thunk help channel the irk that builds in me toward that greedy dark bird with the brown head, parked in the feeder box, placidly grazing, while all the other smaller birds swoop and dart around it. The cowbird is as stubborn as a squirrel: No sooner do I chase it off than it's back again.

I have not liked cowbirds ever since I found out some years ago that they lay their eggs in the nests of smaller birds, forcing the parents to raise an enormous youngster, which often pushes its competition to their tiny, untimely deaths. One hundred to two hundred species of birds—depending on one's source—are parasitized by the cowbird. When I am sufficiently spiritually developed, no doubt I will have more faith that theirs is a necessary part of nature's plan and I will love cowbirds just the same. They really are quite useful, eating up pounds of unpopular insects, and it doesn't seem right that cowbirds are being systematically exterminated in some places to help save threatened species. Still, I don't love cowbirds.

❧ ❧ ❧

ONLY ONE BROWN-HOODED BIRD

The male cowbird is the only black bird with a brown head. Smaller than a robin but bigger than a sparrow, it feeds with its tail cocked up. The female cowbird is dull brown all over and harder to identify. The Brown-headed Cowbird is found all over the United States.

MARCH 28

Song Sparrow

I've been hearing a new song on the block: It begins with three sharp tweets, followed by a slightly higher trill, and finishes with a happy little song, something like *Tweet! Tweet! Tweet! Tril-l-l-l-l! Spring is SO fine!* I'm becoming convinced that many birds are ventriloquists! Although the song is loud and frequently repeated, it took me fifteen minutes this afternoon to find the little brown, streaked singer, perched in plain view at the end of a low, bare branch on the lake-facing bank of the dune. I've never thought of sparrows as impressive singers, but this bird looked very much like a sparrow, and it was singing its heart out.

I sat down on a stump on top of the dune to look up *sparrow* in my bird book, while the bird sang on, head thrown back, stubby little beak a wide V against the deep green-gray of the lake, where gulls soared and mergansers flew by in jetlike squadrons. I confirmed my sparrow I.D. in minutes, but that didn't really get me anywhere: Pictured in my bird guide were close to thirty sparrows, most of them streaky little brown-and-white birds with stubby, conical beaks.

His eye is on the sparrow, my foot, I thought. *Which* sparrow? The Vesper Sparrow? The White-crowned Sparrow? The Lincoln Sparrow? How about the Chipping, Swamp, Song, Grasshopper, Fox or Field Sparrow? Before I had even the slightest idea which of those, if any, God had *Her* eye on, or which, if any, my singer was, my sparrow flew away.

Later, when I learned what to look for, I could spot a Song Sparrow with ease.

🐦 🐦 🐦

SPOTTING SPARROWS

If you see a sparrow (4- to 6-inch, mostly brown-and-white bird with a short, conical beak) early in the spring, look first for a large dark spot centered like a medallion on a *streaky* breast. The Song Sparrow is the only eastern sparrow to have such a spot and is often the first species of sparrow to arrive after winter. (A western race of the

---

NARROW THE SPARROWS

First look at the breast: Is it plain or streaked? There are about eleven sparrows with streaked breasts, while the majority have plain, unstreaked breasts. Second, eliminate sparrows out of your geographical range. This might get you down to six to ten possibilities. Third, notice the head colors and markings. Some sparrows have quite distinctively striped or brightly capped heads. The good news is that the male and female of most sparrow species look alike.

Savannah Sparrow shows a similar central breast spot, as does the larger Fox Sparrow, identified by that and its foxy-red tail.)

MARCH 29

Eastern Phoebe

A friend called this morning to alert me to some Buffleheads she had just seen on a small pond in Saugatuck. I wasted no time getting there: Buffleheads are exquisite little black-and-white ducks that are supposed to be fairly common here, but which, despite considerable pond-hopping, I haven't been able to find. Today was no exception: The only ducks left on the pond when I got there were green-headed Mallards. It was a pretty pond, though, so I sat down to take in the rows of cattails fluffed out from winter, the bright Red-Osier Dogwood behind them and the reflections of the tree branches snaking over the sky-gray surface of the water.

Suddenly I heard a rapid, almost spoken call: *Phoebe? Phoebe? Phoebe?* It didn't sound at all like the high, slow chickadee whistle. This was lower, faster, insistent and seemingly without end. Soon I held a dark gray, white-breasted, slightly crested bird in the lens of my binoculars. I looked it up in my bird book, and what do you know? I had found my

---

THE FLYCATCHER SILHOUETTE

Almost all of the eighteen sparrow-sized (4- to 6¼-inch) flycatchers have strangely pointed heads, somewhat like a titmouse's but blunter. Flycatchers are earth-toned; they're gray, brown, olive or slightly yellow, with smooth, plain-colored breasts. The Eastern Phoebe's dark gray body and white breast are in the sharpest contrast.

---

phoebe! It even wagged its tail, just as my book described. If it weren't for the call, though, I might have had as difficult a time with this small bird as I had had with the sparrows. The phoebe is a flycatcher, and flycatchers appeared to resemble one another even more closely than sparrows do.

ə ə ə

PHOEBE: THE FIRST SPRING FLYCATCHER

If you're a Northerner and see a dark, white-breasted, pointy-headed bird in March or early April, it's probably an Eastern Phoebe. Nobody seems to know why the phoebe, which eats insects, often arrives before insects do. The phoebe seems adaptable, though, eating suet and even seeds until the bugs hatch. Phoebes are usually found near water. The Eastern Phoebe ranges all over eastern North America and western Canada.

MARCH 31

Brown Creeper

I've been looking all winter for Brown Creepers at my suet feeders, because just about every list of winter feeder birds includes them, but I had never seen one. Then today, I was returning from an unusually uneventful walk around the lagoon when I saw a little, speckly-winged, brown bird spiraling up a tree trunk. It was the size and coloring of a female House Finch, but it wasn't behaving like a finch. The only birds I knew that trotted up tree trunks were nuthatches and woodpeckers, but they weren't brown. The bird flew off, though, before I had time to figure out what it was.

Not to worry. I had almost reached my van about fifteen minutes later when I saw another one, again spiraling up a tree. This time I noticed its thin, delicately curved beak, hardly a finch's thick, businesslike mouthpiece, and the white breast, nailing it for sure as a Brown Creeper. As soon as I got home, I saw a third Brown Creeper spiraling up a tree outside my deck. Repetition is a good teacher, and nature is one of the best.

꿈 꿈 꿈

THE SPIRAL BIRD

Not only is the Brown Creeper the only bird that habitually goes up a tree in spirals —it rarely walks down, but drops to the foot of another tree and starts over—but it is also one of the few winter birds with a thin, curved beak seen in the north. (Sometimes the Carolina Wren will winter over, but its tail sticks straight up and it doesn't spiral up tree trunks.)

April

Wood Duck

This morning I was drinking my coffee and watching the treetop view through my cinematic stretch of windows—small birds dashing into the feeders, woodpeckers flashing through the hemlocks, clouds rushing toward and over the house—when I was presented with the oddest sight I've ever seen through those windows: A duck flew by and landed high in a leafless tree, teetering precariously on webbed feet. A duck in a tree? Well, it *was* April Fool's Day. While I focused the binoculars, however, a second duck cruised in and landed on a parallel branch directly below the first. The lower duck stretched its neck to peck a few times at the higher duck's feet, and then both turned toward the lake and sat there for about a half an hour in graceful silhouette.

I had never seen a duck in a tree before. I might not have believed my eyes except that the pair, easily identified, were Wood Ducks, the only native perching duck in North America, an exquisitely beautiful bird named for its unducklike habit of nesting in tree cavities. I sit here amazed that birds as extraordinary as Wood Ducks are to be found in perfectly ordinary environments, in plain sight of a perfectly ordinary person like me. Why have I never seen one of these before? The Wood Duck, although not as commonly seen as some other ducks, is not rare. It is even a legal game bird, found all over the eastern United States, as well as along the West Coast. This was not always so. By the turn of this century, the Wood Duck was almost extinct, victim to lost habitat and uncontrolled hunting. Since then, programs to provide special nesting boxes in large numbers have helped reestablish Wood Duck populations.

It's not hard to believe that such a beautiful bird would be threatened. The male Wood Duck looks too perfect to be real: a scarlet eye, a bright red-and-black bill, an emerald-and-chestnut body with clean-cut, white details; even the huge crest drooping from the back of his head looks carved and painted. The female, although the typical brown of a female duck, wears a white comet around and streaking back from each black eye.

DUCKS AND GEESE WITH WHITE FACE CRESCENTS

Several ducks and geese have a white head-slash, almost always found on the male, that is easy to see and, after species out of range are eliminated, quickly narrows the possibilities:

CANADA GOOSE:
A fat slash behind each eye

GARGANEY:
A large "eyebrow" over each eye

WOOD DUCK:
A thin finger behind and under each eye (male), a back-pointed streak around the eye (female)

BLUE-WINGED TEAL:
A vertical slash in front of each eye

NORTHERN PINTAIL:
A thin line extending from a white breast up the back of each side of the head

BARROW'S GOLDENEYE:
A tear-shaped slash in front of each eye

APRIL   2

Eastern Bluebird

A resident at a Hindu monastery in the nearby town of Ganges told me recently that he'd put up thirty bluebird houses on the ashram's ample grounds. Having gotten permission to check them out, I went there today and roamed the acres of orchards, ponds, fields and woods under a dark, threatening sky. Hundreds of birds sang: robins and blackbirds, Song Sparrows and cardinals. The air smelled verdant.

And I saw bluebirds. The first was perched at the top of a fuzzy

---

INVITE BLUEBIRDS TO YOUR PLACE

For information about making or obtaining bluebird nesting boxes, write to the North American Bluebird Society, P.O. Box 6295, Silver Spring, MD 20906-0295. Include a stamped, self-addressed envelope.

Staghorn Sumac, a deeply iridescent butterfly-blue bird, with peachy sides and a white place at the belly, between the feet. The bluebird wasn't at all shy; it stayed as long as I was willing to stand there holding up my heavy binoculars. Soon I saw the female, too, not quite as blue as the male, but otherwise the same, restless in a gnarly apple tree.

I'd never seen a bluebird before. Sixty or seventy years ago, bluebirds were nearly as numerous as robins, nesting in dead trees, wooden fence posts and telephone poles and bluebird houses. Then their habitat began shrinking. Along came metal posts and poles, as well as aggressive imports like starlings and House Sparrows, and the bluebird faded from common sight; some claim the population fell as much as 90 percent in only fifty years. Today the bluebird is making a comeback, thanks to the thousands of bluebird nesting boxes that are being put up by bluebird lovers all over the country.

ea ea ea

THREE BLUEBIRDS

The blue-bodied, peach-sided, white-bellied Eastern Bluebird is so similar to the Western Bluebird that it's hardly worth arguing about, as their ranges rarely overlap. The Eastern Bluebird is the only bluebird east of the Rockies. The third species, the Mountain Bluebird, does share territory with the Western Bluebird, including most of the western United States and Canada, but the male Mountain Bluebird's breast is blue, not peach-colored.

APRIL   4

Golden-crowned Kinglet

Yesterday at the monastery I discovered a pair of fat little birds dangling off the plum-colored branch ends of blueberry bushes that edged a quiet pond. Round, olive-colored birds with black, white-barred wings, they were no bigger than hummingbirds and squeaked like mice. The pair gave me a good chase, flitting restively from branch end to branch end, but I finally caught the male in my binoculars, and it took my breath away: He was a Golden-crowned Kinglet, no doubt about it. The top of the kinglet's head was brightly striped with orange, yellow and black. The female was similarly crowned, but with yellow outlined in black.

I'd been looking for kinglets along the lakeshore, where my birder friend Joan had sworn she'd seen them. "Listen for high-pitched squeaks," she suggested, warning that kinglets keep moving, making them hard to spot. Once I found them, though, I knew immediately what they were. Nothing looks quite like a kinglet.

❧ ❧ ❧

THE BIRD THAT'S HEARD, BUT NOT OFTEN SEEN

The Golden-crowned Kinglet is not uncommon: It winters all over the United States, even in the snowy North (although it's rare at feeders). To find one, listen for high-pitched squeaks and then look for a tiny ball of olive-colored feathers bouncing from branch end to branch end. In the southern half of the country and in warmer seasons in the North, you might also see a Ruby-crowned Kinglet, a similarly tiny, plump, olive-colored bird, but lacking the bold black crown stripe: The Ruby-crowned Kinglet male's crown is a dab of crimson; the female lacks any bright head color.

APRIL 5

Spring Peeper

Afterword: When Earl Wolf, naturalist, read my manuscript before publication, he suggested that I mention Spring Peepers. "Birds aren't the only spring singers," he said. "Spring Peepers, one of the loudest heralds of spring, begin peeping while the ice is still out there. It's as if they just can't wait." The rhythm of their song, slow at first, becomes faster as things warm up. "They're everywhere," said Earl. "Every wet ditch from here to Florida is making all this noise."

I told Earl that I hadn't included Spring Peepers because I'd never been able to find one, so I was never certain what it was I was hearing. So Earl described some amphibious spring songs I might hear so I could begin sorting them out.

🐸 🐸 🐸

FROG AND TOAD SPRING SINGING

FROG OR TOAD	SPRING SONG
Spring Peeper	A loud, high-pitched peep that can be heard for a mile or more
Chorus Frog	Sounds like a thumb run down a stiff plastic comb
American Toad	A musical, high trill
Fowler's Toad	Sounds like a dying, whining cat

QUICK MOURNING DOVE I.D.

Listen! Do the wings whistle when it flies? Is it crying softly, *Coo, coo, coo?* Look at the tail: Is it long and diamond-shaped, with flashing white edges? That's a Mourning Dove!

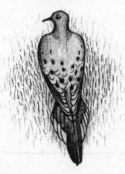

APRIL   6

Mourning Dove

I've been watching a Mourning Dove for nearly a month now. A soft brown bird with black-spotted wings and about a foot long, it started out in the trees quite a bit north of the house but has recently settled in the nearest tree to the deck, flying in for a few minutes each day to feed on cracked corn and sunflower seeds. A friend told me that Mourning Doves mate for life*, so I was beginning to feel sorry for my lonesome dove. I wondered if her mate had died, who knows how long ago, or maybe even had abandoned her. I needn't have worried: Today she was not alone. "She" had become "they," kissing and cooing.

When I was a young girl in Lahore, Pakistan, I slept in a room that opened onto a whitewashed veranda, a hard-packed driveway lined with purple and yellow pansies and a yard filled with trees and birds. Every morning the sound of hornets hovering in the porch corners and the cooing of "morning doves" awakened me as gently as a feather.

🐝 🐝 🐝

HOW TO TELL A PIGEON FROM A DOVE

Only two doves are common to most of the United States and Canada: the Mourning Dove, which is native to North America, and its city cousin, the Rock Dove, imported from Europe and often called a "pigeon." Both have plump bodies about a foot long and small bobbing heads, but the Rock Dove's tail is short and fans out in flight, while

* In fact, although they may mate for several consecutive seasons, Mourning Doves do not always mate for life.

the Mourning Dove's tail is long and diamond-shaped, with flashing white edges. The far South is also frequented by the six-inch Common Ground Dove, the White-winged Dove and the Inca Dove.

APRIL 8

Mute Swan

Sometimes nature is so accommodating that it makes one feel as if an invisible director is calling the takes. Today was one of those days. I was taking a visiting photographer to my favorite lagoon, and it was a glorious day, the first really warm day of spring, the temperature nearing eighty. The noonday light was so bright that Doug decided not to take his camera, which was fine with me, since I've learned that that usually insures you'll see something good. Sure enough, after we'd rounded the last bend, we were greeted by an extraordinary sight: On the opposite shore, across about a hundred yards of water, a large, white swan sat atop an enormous, six-foot nest that looked to me like a recycled muskrat lodge.

"She's spread lily pads all over it!" exclaimed Doug, peering through my binoculars. When it was my turn to look, I noticed that Doug's "lily pads" were creeping slowly toward the swan, who flipped them into the water with her bill when any of them got too close; I counted fifteen turtles sunning themselves on the slope of the swan's huge nest.

The male swan, halfway between ourselves and the nest, suddenly began swimming toward us aggressively. We talked about going back for Doug's camera but worried that the male swan might attack if Doug waded out toward the nest, or maybe the female swan or the turtles would be gone when we got back. "Do it anyway," I suggested. When we returned with the camera about twenty minutes later, a big Canada Goose had appeared. For the next half hour—the time it took to get the picture— the goose nonchalantly strolled farther and farther away from us, drawing off the male swan, who hissed and huffed, puffed and fluffed, at him, feathers curved up and back like angel wings and neck shooting from a drawn-back U to a straight-out shotgun barrel. With unbelievable cool, the goose kept sidling off, always just barely out of reach.

The battle behind us, Doug rolled up his pants and waded into the muck, standing in thigh-deep water where the male swan had been, while I watched from a damp clump of grass. The female swan was still on her nest. There were more turtles than ever adorning it. No sooner had Doug framed the picture, however, than she tucked her head in. I hooted at her, my beseeching calls bouncing back at me across the water, but nothing would induce her to look up.

After ten minutes, Doug came back, squeezed the water out of his pant legs, and we basked in the sunshine awhile, watching the dramatic scene continue behind us, as the swan challenged the goose. At last, the female swan deigned to raise her head, and Doug waded back out and focused in again. The picture was perfect: The nest was now nearly solid with black turtles, sun glinting off their backs; on top lay the brilliant white swan, orange beak dishing off impinging turtles; behind stretched a deep red band of Red-Osier Dogwood; behind that rose two gray-green hillocks, and the sky. I closed my eyes to hold the moment.

"Look!" hissed Doug back at me. In the brief seconds my eyes had been shut, the picture had changed: Three deer had stepped into the top of Doug's frame, directly behind the swan's nest. Three heads turned toward us, before they bolted.

Doug got his picture, hoping the exposure was good. Then we just sat in the sun, stunned by the exquisitely timed appearance of the deer. A muskrat swam by, maybe ten feet away from us. A pair of mergansers flew overhead; numerous Killdeer (see page 218) shrieked and flung themselves hysterically through the air; big fish jumped right out of the water, tail and all, ker-splashing back. A wind came up, rippling across the lagoon; the muskrat moved to a smoother place, out of the wavelets. The sun soaked into our dark sweatshirts. We sat there for a long, long time.

Mute Swans, imported from Europe, do not migrate as our wild swans do, and can be as aggressive as they are beautiful. Fiercely territorial, they often discourage native species. In some areas, at least in Michigan, Mute Swans have multiplied so rapidly that officials have resorted to oiling their eggs, which prevents them from hatching, to help control the population explosion.

HOW TO KNOW A MUTE SWAN FROM A WILD SWAN

All swans look pretty much alike, but you can always tell a Mute Swan by its orange bill. The two North America wild swans—the Whistling Swan and the Trumpeter Swan—have black bills. (See page 283.)

APRIL   10

Dutchman's Breeches

The race is on: Which will be the first spring flower? It all depends on who sees it, perhaps. My friend Mary saw Spring Beauty first (see page 92), and Judy, who runs our local bookstore, swears it was Hepatica. Mine was Dutchman's Breeches, which I found this morning right in my front yard, maybe fifty feet from the lake: delicate, strangely shaped, moon-pale flowers about an inch wide, dangling like tiny bloomers hung up to dry on the stem. Frilly, parsleylike foliage grew lushly underneath. When I walked along the road, I found more of them, often in large patches. After a while, they seemed to me like little butterflies hovering over a lacy green carpet.

🐸 🐸 🐸

DUTCHMAN'S BREECHES OR SQUIRREL CORN?

Squirrel Corn, whose early spring flowers and foliage closely resemble Dutchman's Breeches, has a heart-shaped flower. Dutchman's Breeches are shaped like tiny, delicate, swallowtailed butterflies, or, of course, upside-down breeches. The two plants flower at the same time, often in close proximity, like fraternal, but not identical, twins.

APRIL   11

Spring Beauty

Last night at about 1 A.M. I heard large animals running back and forth across the roof. I leapt out of bed and aimed a flashlight beam on a skinny branch outside the window. The suet ball I'd arduously slipped there off the end of an eight-foot molding strip—which I'd purchased at the lumber yard for just that purpose—was still hanging from its loop of string.

Until now, the raccoons had stolen every hunk of suet I'd hung up, strung up, wired up, or encased in expensive metal brackets, visiting me nightly, even in the rain, leaving the string in the driveway. I finally started taking the suet in at night, returning it in the morning for the woodpeckers, but one afternoon, before my unbelieving eyes, a little black hand reached down from the roof, grabbed the netting around a seeded suet ball and yanked the whole business out of sight!

Everything in order, I went back to bed, telling myself I'd only been wakened by the storm. This morning the world was soaked and misty, horizontal branches pearled with raindrops. Fuzzy white gulls flapped eerily past in the fog. By the time I'd poured my second cup of coffee, I had been visited by chattery juncos and chickadees, the Mourning Doves, cooing and necking, a boisterous jay, quite a large goldfinch, a pair of Purple Finches, a titmouse, a White-breasted Nuthatch, a Red-breasted Nuthatch, a fluffed-out robin perched stoically in a dripping tree, a black squirrel, a cowbird and a wary cardinal, unseen but heard, calling down the chimney. Only the woodpeckers were missing. That's because their suet was missing, too. I'd been fooled again.

I went to Mary's house to check out her claim that Spring Beauties were the first flowers of spring. They were blooming all over the place, simple, half-inch blooms, each of the five pale petals etched with darker pink lines, the small cluster of blooms dangling between a green V of grasslike leaves that emerged halfway up the stem. The lawns that loomed before stately old cottages were thick with them, like strewn rose petals.

ぷ ぷ ぷ

HOW TO TELL EARLY SMALL PINK FLOWERS APART

Even though these early pink wildflowers look quite similar, count the petals. Then look at the leaves.

SPECIES	PETALS	LEAVES
Toothworts	Four	Strawberry or serrated
Spring Beauty	Five	Grasslike V
Trailing Arbutus	Five	Big and leathery
Wild Geranium	Five	Hairy, five-lobed
Wood-sorrel	Five	Cloverlike
Hepatica	Six to ten	Three-lobed

APRIL   12

Cut-leaved Toothwort

Today I discovered a cluster of simple pinkish-white, four-petaled, dime-sized flowers dangling over a whorl of deeply serrated leaves. The whole plant wasn't more than six inches high. When I identified my plant, I discovered that it was a Cut-leaved Toothwort which, along with a number of other toothworts, is called "Wild Horseradish" for its peppery-flavored roots, and toothwort for the roots' toothy appearance. It can grow as high as fifteen inches.

I went back to dig some up. The plant looked nothing like horseradish, a large-leaved plant with big roots. The Cut-leaved Toothwort roots looked like a delicate string of oval beads. They really did taste like horseradish, though.

≈ ≈ ≈

SOME SIMPLE, SMALL, FOUR-PETALED, WHITE OR PINK FLOWERS

Springtime seems to be a popular time for four-petaled, pinkish or white, dime-sized flowers. Many of them are either one of the many kinds of cress or a species of toothwort.

SPRING WILDFLOWER	*LEAF DESCRIPTION*
Cut-leaved Toothwort	Three deeply serrated leaves whorl from the middle of the stem
Toothwort	Three-part leaves, like strawberry leaves
Spring Cress	Bottom leaves smooth-edged, rounded; higher leaves narrower with jagged edges

APRIL   13

Red-necked Grebe

After the hot, sunny day of the swan, temperatures dove below freezing for a couple of days, followed by two days of fog, followed by several warmer days of rain pouring into a churning, gray-green lake and lashing against the windowpanes. Each day, however, the weather would take a little break from whatever it was doing, and I'd hike along the beach with an umbrella, binoculars and a bird book, looping back along the road to look for wildflowers.

Yesterday, as sometimes happens, I was more absorbed by my thoughts than my surroundings, but I hadn't gone more than ten minutes down the rain-beaten beach when I was jerked out of my introspection by two small, strange silhouettes floating close to shore about fifty feet ahead of me. I focused my binoculars, which now accompany me whatever my mood, on the oddest "ducks" that I had ever seen: Two sharp little heads topped long necks, which shot out of stubby duck bodies slightly smaller than Mallards'.

It took me a long time to identify them, not only because they kept vanishing underwater, but also because there was nothing like them pictured in the duck section of my bird book: a male diving duck with a long, red neck, a somewhat triangular head that was half black (top), half white (bottom) and a beak that was long, sharp and yellow. I finally found them in a section called "Grebes" some twenty-five pages away: They were Red-necked Grebes, fairly rare birds here which sometimes migrate through the Great Lakes.

As I walked toward them, the grebes didn't take flight, as mergansers, ducks and gulls usually do, but simply dove and came up farther away. They preceded me down the beach in this manner for about a half a mile. Later, I read an explanation for this: Grebes are wonderful swimmers and divers, but they don't much like to fly, finding takeoff an awkward business. When a grebe feels threatened, it goes down, not up. To migrate, they fly by night, float by day.

ða ða ða

GREBES: THE LONG-NECKED "DUCKS"

Most grebes can be distinguished from ducks by their long, thin necks and their rather pointed heads. Like a loon (see page 139), a grebe dives, eats fish, deals with water better than air, holds its head low in flight and swims with its babies on its back. Some grebes even have a loonlike laugh. Grebes often are pictured very near the beginning of guidebooks, right after the loons. Five of the six species of grebes found in North America frequent the West, especially the Northwest; three are also found in the Atlantic seaboard and Great Lakes states.

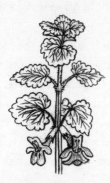

APRIL   14

Ground Ivy

During a quick little dodge out into the nasty day for some fresh air, I found a little woolly plant with minute, purple, snapdragon-lipped flowers and maybe half-inch, rounded, scalloped leaves. The two-inch-high plants grew thickly from long stems that lay along the ground, shooting roots downward and leaves and flowers upward, forming a woolly carpet across the lawn of a still-empty cottage. I was sure it was too insignificant to be mentioned anywhere.

To my surprise, this little plant was given full honors in every one of my flower books. It is Ground Ivy, which, despite having been imported from Europe, has spread all over North America, even into northern Canada. A member of the mint family (see page 290), its leaves are pleasantly aromatic when crushed. In medieval times, Ground Ivy, known also as Gill-Over-the-Ground, was esteemed in Europe as an essential ingredient for the local brew.

ঌ ঌ ঌ

TINY GROUND IVY

Look for Ground Ivy in lawns and gardens. Although it doesn't look like ivy at all, it spreads as ivy does, along ground-hugging stems that root here and there, forming a carpetlike ground cover. A member of the mint family, Ground Ivy's square-shaped stem is thick with quarter-inch (or smaller) purple-lipped flowers and half-inch, rounded, scalloped leaves.

APRIL 15

Lesser Scaup

 At last the sun came out this afternoon and things began to warm up. I got out my Michigan map book, binoculars, guidebooks and warm scarf, and pointed the van inland along country roads, trying to get lost in hopes of discovering something delightful I hadn't known was there. As I drove along, I realized that even breezing along at forty-five miles per hour, I could name most of the evergreens I saw along the road.

By about five o'clock, I found myself at Hutchins Lake, a body of water just barely large enough to be called a lake. I parked the car where the street ended in a little boat ramp and followed an unpleasantly tissue-strewn path around the shore. As I did so, about a hundred little "ducks" bobbed away from me on the water. Soon I came upon a shack that had been trashed and graffiti-ed and was missing its entire front wall. I could hear all sorts of birds in the damp woods, so I sat down on the exposed shack floor to become part of the scene.

The ducks slowly floated back, along with one pair of Mallards that looked big as geese next to the tiny paddlers. I could take my eyes, practically glued to the binoculars, off them only long enough to look them up in my bird book. I guessed them to be the Lesser Scaups pictured there, and sure enough, occasionally one would make a hollow *scaup* sound. About sixteen inches long, the males had velvet purple-black heads, round, brownish-gray bodies and chalk-white bills with a black tip; the females were evenly brown with a little white patch around the bill. I thought I must be in Australia.

ઢ ઢ ઢ

HELP WITH SCAUP LOOK-ALIKES

Except for the Tufted Duck, which prefers salt water, scaups and similar-looking ducks are found in most of the United States. Once you've found a scaup-looking bird— small duck with purple-black head, black chest, white underparts and gray wings—look at the bill:

SCAUPLIKE DUCK	BILL
Lesser Scaup	White
Greater Scaup	White
Ring-necked Duck	Black with white ring
Tufted Duck	Black

APRIL   16

Belted Kingfisher

Yesterday, while I was watching the scaups at Hutchins Lake, a blue-and-white Belted King-fisher flew right into the middle of my binocular focus! For a split second I took it for a Blue Jay, but the ragged crest spraying off its head and the enormous beak told me otherwise.

The kingfisher landed on a branch of a tree that hung over a quiet section of the lake, from time to time piercing the silence with a loud, rattling call. An answering rattle helped me locate another kingfisher perched on another lakeside branch. The two were wonderful to watch, leaving their perches suddenly to dive, with impressive splashes, headfirst into the water, sometimes making several attempts before returning noisily to the branches. The two birds looked almost exactly alike—dull blue head, wings and tail, with white underparts banded across the breast with blue —except that the female had a rusty-colored band below the blue band.

🐸 🐸 🐸

ONLY ONE COMMON KINGFISHER

Big and loud, the Belted Kingfisher is easy to spot: *ke-ke-ke-ke-ke!* Although only an inch longer than a Blue Jay, the Belted Kingfisher looks much bigger: heavier, with a sharp, finger-length beak. An early migrator, moving in ahead of time to pick out its fishing grounds and defend them vigorously, the Belted Kingfisher is common to lakes and rivers all over the United States and Canada.

APRIL   17

Terns

This morning I woke just before dawn and was about to tuck back in when I remembered that dedicated birders get up at that hour, claiming that's when birding is the best. I thought I'd see if this was true. It was cold on the beach, under forty degrees, but the sky was clear and blue. I walked southward, sidestepping little waves that lapped in every two seconds. The lake stretched to the horizon on my right, the sun was rising but still hidden behind the cottaged dunes on my left, and an amazing number of birds were flying toward me, going north.

Here is what I saw in half an hour: hundreds of Red-breasted Mergansers, flapping their wings hard—no gliding—and low over the water; a pair of Blue Herons, necks crooked back in an S; a pair of honking Canadian Geese; three pairs of Mallards, each at a different place on the beach; many flocks of small black birds flying frantically over the spiky dune treetops; a flock of male Red-wing Blackbirds, with sunlit epaulets; hundreds of Herring Gulls.

I haven't paid much attention to gulls, thinking that a gull was a gull around here, but I noticed three awfully small gulls flying behind one group of Herring Gulls, and through my binoculars I saw three elegant, swallow-tailed birds. I knew they were terns from their black caps; my cousin Richard had pointed this feature out to me when I was in Georgia. But which tern? There were eighteen terns in my bird book, and most looked alike to me. I narrowed them down to the four that frequented this end of the Great Lakes, only three of which had black caps: the Caspian, the Common and the Forester's Tern. That was as close as I got. Surely it was triumph enough to be able to tell a tern from a gull, for a person who'd never even heard of a tern before this year.

&a. &a. &a.

HOW TO TELL A TERN FROM A GULL

If a white, gull-like bird has a black cap, it's a tern. Not all terns have black caps: Two are black-bodied almost to the tail, and the capped species lose most or all of the cap after breeding season. Still, although some gulls have black heads, none have black caps. Most terns have forked tails, graceful, pointed wings and are smaller than gulls. Also unlike gulls, which usually thieve their fish, terns are fishing birds, often hovering, then diving straight into the drink from an impressive height.

APRIL   19

Bonaparte's Gull

About nine o'clock this morning, I sat at the top of my beach stairs, elbows braced on my knees, gazing through steadied binoculars at an enormous flock of crested Red-breasted Mergansers. In only a few minutes, thousands of birds had flapped left to right through my round field of vision, which was focused on a section of lake between the horizon and the beach. Despite their vast numbers, however, they would have been hard to see without binoculars. Skimming the waves with businesslike wing beats, they became black dots camouflaged against a dark, choppy lake.

The only birds I saw against the sky were gulls, which I ignored until five very small ones with black heads flew lightly into my circular field. They were as petite and elegant as terns, with a tern's bouncy flight. I followed them for a full minute, noticing the slender, black bills and the black-tipped, gray-topped wings, before they vanished north. There were five species of gulls with black heads in my guide, two appearing here. I guessed my birds to be the more common Bonaparte's Gull, named not for Napoleon, but for avian taxonomist Charles Bonaparte (1803–1857). About half the size of our Herring Gull, I'd have taken it for a tern if I hadn't discovered only yesterday that just as no gull has a black cap, no tern has a black hooded head.

WHY IT'S SO HARD TO IDENTIFY GULLS AND TERNS

My bird guide pictured twenty-one variations on six black-headed gulls. No wonder differentiating between the thirty species of gulls and terns in North America is such a challenge! Not only are there many close adult look-alikes, but gulls and terns take two to three years to reach adult plumage, changing appearance each year, sometimes quite radically. To further complicate matters, some of the easiest-to-spot field marks are only temporary: The black cap on the tern, for example, as well as the black head on a black-headed gull, may disappear after breeding season. There is some good news: Males and females usually look alike.

ð ð ð

BLACK-HEADED GULLS

Black-headed gulls are quite small, the biggest being the Laughing Gull (13 inches) and the smallest the fairly rare Little Gull (9 inches). (A Herring Gull is 20 to 24 inches.) Their black heads appear in spring, fading or turning white by fall. To tell black-headed gulls apart, try to notice the color of the feet and the beak. Although these gulls closely resemble one another, only the first two overlap territories.

ADULT BLACK-HEADED GULL	FEET	BEAK
Bonaparte's Gull (Continental United States)	Orange	Black
Little Gull (East Coast, Great Lakes)	Orange	Black
Laughing Gull (East Coast)	Black	Orange
Sabine's Gull (West Coast)	Black	Black, yellow tip
Franklin's Gull (Inland United States)	Orange	Orange
Common Black-headed Gull (East Coast)	Orange	Orange

APRIL 20

Rock Dove

About a week ago, I saw a fat, city-type, pink-footed, iridescently feathered pigeon perched on a tree outside the deck. The following day, the pigeon had arrived on the deck, and I had to admit that it looked rather pretty out there, pecking away with the squirrels. I told my neighbor Mary that I had a pigeon at my house, and she didn't believe me. "Are you sure it isn't a Mourning Dove?" she asked. None of my other friends had ever seen a pigeon around here either. No one knew what it was doing there. It's hard to believe that it could ever be considered a rare bird, but a pigeon in the country can be as startling as a pheasant in New York City.

🐦 🐦 🐦

PIGEON OR DOVE?

Our urban "pigeon" is not really a pigeon but a Rock Dove, imported from Europe. Once a cliff-dweller, it took quickly to city living and tall buildings. The only fairly numerous North American pigeon is the Band-tailed Pigeon, a gray bird with a wide white band across the end of its tail, found in parts of the Southwest and along the North American West Coast. Our other pigeon species tend to be extinct (the Passenger Pigeon) or occupy tiny southeastern territories (the Red-billed Pigeon and the White-crowned Pigeon).

APRIL   22

Painted Turtle

This afternoon, about four minutes' drive from here, I spotted a little pond I hadn't seen before, parked the van across the road and sat down on some moss by the water. Nearby, some kind of sparrow with a chestnut head hopped between the willows along the shore. The pond fanned out, big as half a city block, choked with fallen logs, tangled gray willows and hundreds of dark green horsetail whips (see page 104). Nothing moved much.

It felt delicious to be sitting by a sunny pond only minutes from my house, so different a place from the lakeshore. A pair of cardinals called to each other, one right over my head: *Birdy birdy birdy birdy birdy!* Blood red flashed across glossy black water. After half an hour, I took one last look through my binoculars: Two big black turtles were sunning on a log! I looked further and found three more, at the far end of the pond, sun patches white on their plate-sized, black, oval backs. How had I missed them? They looked like the turtles on the swan's nest at the lagoon, but although I was a little closer to these, I still couldn't tell for sure what they were.

Back home with my turtle book, I tried to identify the turtles. It shouldn't really have been too difficult—there are only nine turtles native to Michigan, with a tenth variety released from pet shops—but I hadn't been able to get close enough to the turtles to pick up any markings. I thought these had to be a common kind of turtle, because there were so many of them. In the end, I called the naturalist at a state park north of here, Earl Wolf, who guessed from my description—oval, slim, almost black shells—that they were Painted Turtles, even though the books insist that Painted don't get much bigger than seven inches. "Painted Turtles are baskers," said Earl. "And they often look bigger from a distance than they really are."

It's a lot easier to identify a turtle if you can see one up close, but turtles are shy and can vanish for long periods of time (some sources claim that a turtle can hold its breath for up to ten minutes!). If we're not careful, turtles may vanish for good.

&ae &ae &ae

THE VANISHING TURTLE

Until recently (1988 in Michigan), there have been few laws protecting turtles, and their numbers are rapidly diminishing. Kids, collectors and the pet store business are threatening turtledom. Be kind to turtles: Leave them where you see them, and if you see one threatened, save it if you can. I've rescued several from certain highway death. Be careful, though. A Snapping Turtle does not tend to be grateful. (See page 174.)

APRIL 23

Rough Horsetail

I am learning that there is often more in a scene than first meets the eye. Staying put at yesterday's quiet pond, where nothing appeared to be happening, revealed well-camouflaged inhabitants and invited new ones: Before I left, a brown Mallard mama slipped into the pond, and behind her, a shining muskrat. The muskrat glided as far as the middle, saw me and immediately sank. Around the edge grew hundreds of slender, green switches which had joints every few inches that popped when I pulled on them hard. They were so tough that I was amazed to find that the sample I brought home was, of all things, a fern.

I think of ferns as the hypochondriacs of the plant world: Could anything be more delicate, more demanding than a fern? Not all ferns, apparently. This fern was a Rough Horsetail, the cockroach of the fern family: very primitive and very, very tough. Three hundred million years ago, these quarter-inch, four-foot whips grew tall as trees, which by now have probably turned into coal.

ᶓ ᶓ ᶓ

SCOURING RUSHES

There are at least ten varieties of horsetails in North America, many of them common. Some branch out at the joints like umbrella ribs, some don't; some grow as high as fifteen feet, others only six inches. All of them are nicknamed "Scouring Rushes," because they contain silicon, which makes them excellent pot scrubbers. After some experimentation, I found that the little, frilly Field Horsetails, which appear to thrive just about anywhere, but especially in shady, damp places, worked best. (You can often distinguish a Horsetail from other plants by its rough feel.)

APRIL 25

White-crowned Sparrow

I saw the most amazing sparrows pecking at the ground around Mary's feeder today, their bold black-and-white zebra-crowned heads so distinctive that I identified them in a flash. They were White-crowned Sparrows, and seeing them made me realize that sparrows aren't the dull little brown birds I've always thought they were. Having seen a White-crowned Sparrow, I have to make room for *elegant* in my sparrow vocabulary. You have a good chance of seeing one, too: White-crowned Sparrows winter in all but the most northern states, summering in the Northwest and in the far, far North of Canada.

ᶓ ᶓ ᶓ

WHITE-CROWNED VS. WHITE-THROATED SPARROWS

There's another sparrow with a black-and-white striped head: the White-throated Sparrow. White-throated Sparrows, however, have a small but bright yellow patch in front of each eye and a fluffy white throat patch.

APRIL 26

Great White Trillium

Today the world was roaring with bumble bee queens (see page 261), and no wonder. As if by universal agreement, millions of dandelions have suddenly burst into brazen glory in any yard with a lawn. Yards ground-covered in myrtle are thick with periwinkle blue, pinwheel flowers, above which bloom hundreds of daffodils under the graceful, golden sprays of Forsythia bushes.

I walked past all this yellow and blue on my way to the woods today, where, although the layered look continued, the colors changed. The slopes on both sides of the soft, high, wooded dunes I walked between were carpeted in the frilly foliage and delicate pale flowers of Dutchman's Breeches and Squirrel Corn. A foot above this were scattered thousands of the most beautiful white, lilylike flowers I have ever seen: Great White Trillium. I couldn't believe that any wildflower could be so beautiful and still so profuse. Each three-petaled flower bloomed big as a wild rose between three slim sepals (those outer leaflike structures that enclose the bud), atop a whorl of three large leaves on a single stem. In front of me, beside me, on the hills above my head, all around me, the forest was lavishly decorated in lacy green and cream, as if for a faerie queen's wedding.

About thirty kinds of trillium, members of the Lily family, bloom in North America. Most of them are found in the eastern United States and Canada, but at least one, the Coast Trillium, is also found in the West.

PLEASE DON'T PICK THE TRILLIUM!

Before you decide that just one less trillium will hardly make a difference, remember that a trillium plant takes six years to mature from seed to flowering. If you break the stem below the leaves, most trillium plants will die. Trillium is protected by the state of Michigan and may be in your state, too. Before you buy it in a garden shop, be sure it was raised by the nursery and not collected from the wild.

APRIL 27

Northern Harrier

Yesterday, while I was trying to take in the beauty of the flower-filled woods, I saw another flash of white, this one above my head. A hawk flew past, across an opening in the trees, and vanished in front of me behind a dune. I was just turning to the hawk pages in my bird book when I heard three sharp screams overhead. The hawk was back, this time with a large rodent in its claws. I thought it was the rodent screaming, but I've since read that when a male Northern Harrier makes a kill, it sometimes screams to announce its success to the female. The hawk passed over more slowly this time, in the direction from which it had first come, and vanished.

I found the hawk section of my bird book most confusing. Not only were females and immature hawks often differently colored from adult males, but some hawks had a light and dark morph. There are, for example, dark-colored Red-tailed Hawks and light-colored Red-tailed Hawks (see page 222). To make matters more difficult, I've never gotten more than a fleeting glimpse of a hawk, unless it is soaring very high, very far away. My eyes just aren't trained to differentiate between wing shapes, which appear to be important in the identification and categorization of hawks.

I did notice, though, that the hawk I saw had a long, thin tail, a pearly, fairly slim body, and "fingered" wings. The only hawk I could find that fit that description was the Northern Harrier, common throughout the United States. Sometimes just noticing a few points is enough.

BEGINNING HAWK POINTERS

There's rarely time to check a soaring hawk against bird book illustrations, so you'll want to spend your split second noticing things: Look at the tail. Is it fan-shaped, or long and thin? Is it banded (how many bands?) or plain? If there's still time, check out the underbody colors. Is the body a different color from the underwings? Speckled or plain? Any stripes, patterns, bands or blotches? Even remembering these apparently simple details takes practice.

APRIL 28

Sketchbook

Spring must be here. My friend Bette complained to me just last week that the juncos were taking their sweet time leaving this year. Spring did not truly arrive for her, she said, until the juncos were gone. Today I saw no flash of junco fantail white in the woods, nor did their winter hordes appear at my feeder.

Several days ago, I bought a large sketchbook and began drawing birds and squirrels on the deck. The next day I drew flowers. I'm no artist, so I worried about being correct. I started with tight, timid, barely visible pencil sketches. The next day, encouraged by my artist friends to "Loosen up!" and "Be bold!", I went over all my scared little lines with ink lines and markers. I started brightening things with colored pencils. Wow. Things were beginning to bounce. As I drew, I learned details I'd never noticed before: where a bird's legs meet its body; the backward crook of a bird knee; the varying ways leaves attach themselves to a stem; where the blue is on a jay; the shape of a squirrel's ear.

Later, I went back again and tried to make the random sketches on each page into a whole drawing. I drew in more things from memory, added background. By this time I'd stopped caring about relative size or accuracy. Each page unified into a particular day of surprises. The result was strangely moving. Even though my drawings were childish, unintended for artistic or scientific value, they jumped with life.

🐦 🐦 🐦

A NEW WAY TO SEE

Draw something and you will never see it the same way again. I recommend sketching anything you care about. Don't worry about your artistic talent; the point is to learn to see in new ways. Drawing is especially valuable to those of us who haven't done much of it, the process more startling to us than it is to those who have already developed this way of seeing. Use any paper, any drawing materials: pencils, pens, markers, crayons, paints, whatever. A reassuring book for the nervous nature sketcher is *The Zen of Seeing*, by Frederick Franck (Random House, 1973, paperback).

APRIL 29

Rufous-sided Towhee

It wasn't easy to pick new bird songs out of today's Blue Jay screeches, the chainsaw screams of the flickers, woodpecker percussion, robin carols and cardinal burbling. On a side street several blocks in from the lake, however, one song came through loud and clear: *Drink your t-e-e-e-a. Drink your t-e-e-e-a.* It rang through the air from a very high place and continued until I had tracked it to the tip-top of a huge, droop-branched Norway Spruce. The singer looked very much like a robin to me, but there were some startling differences: It had a black hood that dipped, biblike, below its beak; its back and tail were mostly black; and the middle of its breast was white, with rufous (robin-red) sides. Although I'd never seen one before, the Rufous-breasted Towhee is common throughout the United States. (The back of the western version, or "race," is speckled with white.)

૨ઢ ૨ઢ ૨ઢ

SOME COMMON BLACK-HOODED SONGBIRDS

A black hood is easy to spot and usually found on songbirds smaller than a robin (9 to 11 inches) but bigger than a House Sparrow (5 inches). Most black-hooded birds can be told apart by their startling breast color:

BLACK-HOODED BIRD	APPROXIMATE SIZE	BREAST COLOR
Rufous-sided Towhee	7 to 8 inches	White with rust-orange sides
Dark-eyed Junco	5 inches	Slate (Slate-colored race)
		Black (Oregon race)
Rose-breasted Grosbeak	7 to 8 inches	White with rose bib
Black-headed Grosbeak	7 inches	Orange-yellow
Baltimore Oriole	7 inches	Bright orange
Orchard Oriole	6 inches	Brick red

APRIL 30

Eastern Cottontail

An hour ago, I walked out the front door to have a stretch, and across the road was a rabbit, nibbling around some daffodils, a classic springtime scene. The sun was getting low, the light was glowing, the air nippy. Robins hopped by, unafraid; cardinals flitted over the grass; finches chattered in a Norway Spruce; starlings whistled in tenderly leafing maples. The rabbit noticed me and hopped casually away, with long periods of standing still or nibbling while facing me full on. Its straight-up ears seemed pleated, so that the backs touched each other, the insides opening out on each side like elongated satellite dishes.

At first I thought I'd never identify that rabbit. The six rabbits on the same color plate in my mammal guide looked exactly alike: all rusty-gray. When I flipped to the back of the book to check the range maps, however, I discovered that my anxious comparisons had been unnecessary: Five of the six rabbits don't live here. By elimination I discovered that my neighborhood rabbit was an Eastern Cottontail. It even had the big rusty patch between the shoulder blades that the others don't. Then and there I took time to highlight my local species in all my guidebooks.

HOW TO MAKE NATURE GUIDES EASIER TO USE

It can be helpful to take time to highlight local species with a bright marker next to the species picture. This can be a great introduction to—or review of—wildlife in your area and, by eliminating a great many crazy-making look-alikes, makes field I.D. quicker and easier. It's especially useful for field guides that locate information on a species in two places. (Some field guides put range maps in the back; others put the photographs in front, and descriptive material in back.)

May

Common Blue Violet

On May Day, I used to fill handmade paper baskets with violets, set them at the doors of my favorite grown-ups, ring the doorbell and hide so I could watch their faces when they beheld my gift. I've always loved violets. So far I've found four kinds along or just off Lake Shore Drive. The best patch is blooming wildly down the middle of a little two-track road that cuts between two still-unoccupied frame cottages to the golf course. I counted more than fifty inch-wide, passionately blue blossoms per square foot. Lush with heart-shaped leaves, the violets brought startling color to the tiny, greening lane.

I think they were Common Blue Violets, but I had a hard time telling. Violets hybridize enthusiastically; there are sixty to eighty species in North America alone (five to nine hundred in the world), depending on which authority you read. Some species have odd leaf shapes. For example, Bird's Foot Violet's narrow leaves are shaped like its name, instead of like hearts, as most violet leaves are. Flower color doesn't help much: Violets are not always violet, but vary from many shades of purple, blue and magenta to yellow and white. I've decided to be content with knowing which of the major two violet groups most species belong to (see below).

I've learned some remarkable things about violets. Bees flip over them. The bottom petal of most violets is marked with color, nectar guides leading to a fuzzy "beard" which provides the bee with a foothold. To reach the nectar in the petal's inner spur, the bee must turn upside down, a bit of acrobatics that facilitates pollination.

As profusely as violets seem to grow, most don't start blooming until their fourth year. After the dramatic spring display, they may produce small flowers that remain closed, self-pollinate and produce a prodigious amount of seeds. When dry, the seed pods open and explode their contents as far as four feet—fifteen feet for the Smooth Yellow Violet!

☙ ☙ ☙

STEMMED AND STEMLESS VIOLETS

For all the confusion, violets do fall easily into two main groups: those with leaves on the flower stem (*stemmed*) and those without (*stemless*). A common stemmed violet is the Downy Yellow Violet, with a yellow flower atop each leafy stem and soft hairs on its leaves. The Common Blue Violet is stemless: Each leaf and each flower stem grows directly from the ground.

MAY 3

Mustards

Under the newly leafing trees, beneath apple trees petaled pink and white, mustard-yellow flowers spread, profuse and delicate. I guessed that they were mustard, and I was right, but it was news to me that not only are there many kinds of mustard, but not all mustards are yellow. Many members of the mustard family, like Garlic Mustard, Water Cress and Lyre-leafed Rock Cress, flower white. There are so many kinds of mustard that I feel quite satisfied to call a mustard a mustard without worrying too much which kind it is. Not all mustard family members fit the description below—there are some, for example, that flower along the stem, others with rounded seed pods—but any plant that does pass the mustard test is a good candidate.

☙ ☙ ☙

THREE EARLY MUSTARDS

The first three members of the mustard family that I've seen this year passed the mustard test. Distinguish Garlic Mustard by white flowers and the crushed leaves that smell lightly of garlic; Lyre-leafed Rock Cress by white flowers and slender, smooth-

---

A MUSTARD TEST

On top of the stem look for tiny, four-petaled yellow or white flowers blooming in rounded clusters; below these, long, narrow seed pods; below them, alternating leaves. Although not all mustards fit this description, a plant that does is probably some kind of mustard.

---

edged leaves; Yellow Rocket by yellow flowers and leaves that resemble a head with several sets of little arms beneath, the bottom arms clasping the stem.

MAY 4

Large-flowered Bellwort

I treated Ellen and her Ann Arbor guest to a wildflower walk in the woods today, and nature cooperated by giving us a good show: The trillium was thicker than ever, blooming with huge abandon in breathtaking spreads up and down the dunes and across the forest floor. As if on cue, two deer trotted through the trillium, not a hundred feet in front of us, pausing, ankle-deep in blossoms, to stare at us before dashing up the dune.

We'd all seen trillium blossoms before, but none of us recognized the limp-looking yellow flowers that crowded among them. I had brought along a wildflower book and easily identified the lanky lilies as Large-flowered Bellworts, wilted-looking flowers that might have been almost as large as trillium if the six slender petals actually opened, but they didn't and they never would. Large-flowered Bellwort—an awkward name for such a slender plant—never loses its droop. The top flops over, and the flower stares, hangdog, at the ground; even the large, smooth leaves sag.

꙳ ꙳ ꙳

THREE YELLOW SPRING LILIES

These three yellow lilies are common in springtime. All are six-petaled, but they are easy to tell apart:

Large-flowered Bellwort	(1 to 2 inches): Droops; leaves grow from the stem.
Trout Lily	(¾ to 1½ inches): Petals curl back, like an Easter Lily. A pair of mottled green-and-maroon leaves grow from the base.
Clintonia	(¾ inches): Flowers are small, several on a leafless stem that emerges from a cup of large, shining *basal* leaves.

MAY 5

Sensitive Fern

Today I went over to see my friend Marcia, who permits a bat to live above her ceiling and have free rein of the house. She doesn't even kill kitchen mice but traps and releases them in her woods. We soon were walking through woods, on damp ground thick with low, blooming straw-berry plants and small, green, red-stemmed ferns. "Those are Sensitive Ferns," said Marcia. "I looked them up. But I don't know why they're called 'Sensitive.' They don't seem especially sensitive to me."

I never found the answer to that question either.

꙳ ꙳ ꙳

SEEING THE SENSITIVE FERN

Applying the above process to the Sensitive Fern: It is *once-cut*, meaning that the leaflets aren't lacy, or deeply cut; the frond is almost triangular, largest at the *base*; the *stem*

IF ALL FERNS LOOK ALIKE . . .

To begin *seeing* ferns, first check the *cut:* Is it a once-cut, a twice-cut or a thrice-cut fern (see page 31)? Then look at the *base:* Is it broadest at the base, or semitapered, or tapered? Check the *stalk* for color and length. Finally, look for anything unusual. This process doesn't always identify a fern, but sometimes it helps.

---

is red when young, chestnut when mature; the Sensitive Fern's *opposite* leaflets have an unusual network of tiny veins like dragonfly wings. (So does the Netted Chain Fern, but its leaflets are *alternate*.)

MAY 6

Magnolia Warbler

Today I discovered a whole new batch of birds flitting between six or seven towering spruces just a block or two inland. At first I assumed they were a flock of House Finches. As I watched, however, I couldn't identify most of the birds I saw. There was some sort of chestnut-headed sparrow; lots of kinglets squeaking around the ends of branches; some kind of gray, robin-sized bird and many others that flitted into thickly needled branches before I could catch them in my binoculars.

The bird that finally held still for me was the loveliest bird I've seen yet. As it sunned on a branch end, I practiced noticing details: It was small but not tiny; the body was brilliant yellow with bold, black streaks; a bright band of white slashed across black wings and a black tail; the eye was zebra-striped white over black. It was a Magnolia Warbler. I felt quite smug. I had identified this bird out of almost fifty warblers pictured in my bird book, most of which are more or less yellow. Identifying warblers is not for sissies.

The Magnolia Warbler, an eastern United States bird, reaching into

---

WARBLERS DON'T ALWAYS WARBLE

Another cherished notion hits the dust. Sparrows would give warblers a run for their money in a sing-off; many warblers do little more than squeak or tweet. Some warblers do have lovely songs, but they're just not the North American nightingales I was expecting.

---

western Canada, was named in 1800 by an ornithologist who discovered it in a magnolia tree. Actually, it couldn't care less about magnolias, preferring conifers, particularly spruce, which is where I first saw it.

ða ða ða

"SPRING" AND "FALL" WARBLERS

Until today, I thought the "spring" and "fall" warblers referred to by my bird guide were different species. Not so. Unlike the leopard, which cannot change its spots, the warbler can. Like the goldfinch, the brilliant spring warbler dulls in fall and winter, often losing distinctive field marks (especially the male, which is usually brighter than the female). Few species of birds look more alike than warblers in the fall and winter. If you're going to fool with warblers, start in spring.

WHERE IS A BIRD'S "RUMP"?

If the body had not been so yellow, my Magnolia Warbler could have been a Yellow-rumped Warbler, distinguished from the Magnolia Warbler by its yellow crown. Both birds have yellow rumps. A bird's rump is the patch just *above* the tail (not just below).

MAY 7

Woodchuck

It was a warm sunny day at last, temperature in the seventies, and I walked down the lakeshore road trying to sort out a new outburst of wild-flowers and watch a swarm of swallows swooping and looping just offshore, moving too fast for me to identify. Some distance away, I noticed my neighbor Marti approach, then suddenly stop: Between us, motionless in the middle of the road, sat what looked like a fourteen-pound gerbil. I raised my binoculars as the brown animal scurried away on short legs, disappearing between some shrubs lining an elegant house.

"What was *that*?" asked Marti when we met. I didn't know. I once chased a Marmot that looked just like it from a garage in Colorado, but I didn't have a mammal guide with me to be sure. The onset of warm weather was posing a problem for me: My big down jacket pockets held three guides at a time, but my jeans held only one, usually a bird guide, in a back pocket. When I got home, it only took a minute to identify the furry animal as a woodchuck. It could have been a Marmot, but there was a very good reason why it wasn't.

🐾 🐾 🐾

WOODCHUCK OR MARMOT?

A woodchuck and a Marmot look almost alike, a woodchuck being a kind of Marmot: rabbit-sized, beaver-shaped bodies, with long-haired tails. Tell them apart by range: Woodchucks live east of the Rockies and in Canada, while Marmots—two kinds—live in the Rockies and west. Marmot and woodchuck ranges don't overlap much.

WHEN IS A GROUNDHOG A WOODCHUCK?

Always. "Groundhog" is just a nickname once-removed for "woodchuck," which is our nickname for *Marmota monax*. It's actually a woodchuck that supposedly looks for its shadow. (So how much ground will a groundhog hog if a groundhog could hog ground?)

MAY 8

Black Tern

Last night I was treated to my favorite storm, the kind with belly-felt thunder, horizontal lightning, torrential rain and no wind. The rain fell straight as string. This morning dawned still and clear. By seven, I was walking in the woods, which smelled sweet and loamy, listening. As I stole silently along the sandy path, two birds, hidden in the thick conifers, fluted to each other. I've never heard such a sound before, and it went on and on, back and forth, extraordinary three- or four-note calls that cast a spell on the woods.

When I reached the beach, I found it magic also, strewn with beach glass, thick, old shards worn smooth and opaque by sand and waves. I don't collect beach glass, but today it was irresistible, shining like jewels in the morning sun; emerald, root beer, white quartz, pale sapphire, even the rare cobalt blue. I picked up a pocketful in half an hour.

Now and then, it occurred to me to look up. This is what I saw: some kind of hawk, white underneath and dark on top, with a black streak across the eye, cruising the close-in waters, occasionally hovering; a lone merganser and three kinds of terns—a small, black-capped tern with a thin, orange bill (maybe a Common Tern); a big, black-capped tern with a heavy, scarlet bill (probably a Caspian Tern) and an all-black tern (definitely a Black Tern). I named the easiest.

ᕚ ᕚ ᕚ

ONLY ONE BLACK-BODIED TERN

Found on fresh water all over the United States and southern Canada, the Black Tern is the only common tern or gull with a black body. Its gray wings and black bill are not found on other quite rare black-bodied terns (Black Noddy in Florida or an occasional White-winged Black Tern along the East Coast).

MAY 9

Gray Catbird

A Gray Catbird thonked itself unconscious on Mary's bedroom window this morning, just before I arrived to pick her up. It lay feet-up on the walk to the beach stairs. It isn't often I can hunker down and inspect a bird as if it were a wildflower, while comparing it to pictures in my guide. This one would have been quick to identify anyway, being the only black-capped, gray bird in the book. When we returned from a downtown breakfast, the catbird was perched on a low branch of a bush that held a cardinal's nest containing one egg. I went to get my sketchbook, but when I got back, the catbird was gone. Mary told me that catbirds really do sound like cats, but this one wasn't talking.

I later made a discovery that amazed me: The catbird is not only an apt mimic and ventriloquist, but one of our sweetest songsters. The very next day, I found a pair of catbirds by following an oddly Song Sparrow-like call. In June, I heard a loud meowing in some shrubs and, looking for a cat, found a catbird. In July, I was so astonished by a long, varied and musical bird song that I bought a tape of bird calls to find out what was singing: It was a catbird.

❧ ❧ ❧

ONE BLACK-CROWNED, GRAY BIRD

The catbird is our only big (robin-sized), gray bird with a black-crowned head. Look also for rust-colored feathers under its long, flicking tail. The Gray Catbird is found almost everywhere except the Pacific states.

MAY 10

Rose-breasted Grosbeak

When I moved into this lovely apartment on the lakeshore, I agreed to vacate it from late May to September to make room for summer rentors, who willingly ante up ten times the rent I am presently paying. I wondered if this would be a terrible hardship for me, moving all my things into the garage, finding somewhere else to go for a few months, moving back in come fall. "Do it anyway!" I told myself, my three favorite words after "I love you." I had dreamed for years of living within sight of Lake Michigan, and if this was the price I had to pay, then, I decided finally, I would pay it.

Finding something to do for a whole summer turned out to be a wonderful opportunity to make more dreams come true. A few weeks ago I decided that I would travel around the Great Lakes for the three months, live in my van and later write a book about the experience.* The combined stress and excitement, not to mention the packing, that often precede an adventure involving a large financial and professional risk have been waking me up at six in the morning, leading to the solving of one of life's great mysteries: why those batty birders hit the woods at dawn.

This morning I stepped out into warm, orange air thick with bird noise: A robin grace-noted up and down the scale; cardinals *birdie*d; jays screamed; flickers screeched; finches chattered; Song Sparrows trilled; Mourning Doves cooed; crows taunted; titmice whistled; chickadees *chick-adee*d. The music was voluminous, symphonic. There obviously had been many new spring arrivals, and I began to appreciate how much it helped to know even a few birdcalls.

By following unusual bird songs, I located so many beautiful birds that I don't know which one to write about. I began following a harsh, squirrel-like call, but I couldn't find the bird, even though I stood under the very tree from which the loud sound came. Minutes later, I heard the

---

* It is not this but the following summer, spent in a one-room cabin on a Lake Michigan island (another dream), that is reflected in this book, however (see page 134).

same call in another tree, and this time I found two raucous Red-headed Woodpeckers in the same tree.

Five minutes later, I heard a lyrical song in a tree on the lake side of the road and soon discovered a large black-headed, white-bellied bird with a rose-red breast. It was a Rose-breasted Grosbeak, a bird so lovely it made the cover of one of my bird guides. Soon another male flew in, and then a buffy-brown female. I eventually left them still nipping at the leaf buds.

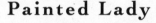

THE ONE AND ONLY ROSE-BREASTED GROSBEAK

The dramatic splash of scarlet below the all-black head of the Rose-breasted Grosbeak is an instant giveaway: Only the Ruby-throated Hummingbird resembles this arrangement of dark top, white underneath and red throat, and not even a beginner is likely to confuse a hummingbird with the parrotlike Rose-breasted Grosbeak.

MAY 12

Painted Lady

All day yesterday and most of last night, vicious winds have been ripping off the lake and tearing into the dunes, just now delicate with spring. This morning I could hardly see out of my twelve west-facing windows, they were so plastered with shreds of tender new leaves. It was the first sign of how fierce the day-and-night-long storm had been. Later, I found Mary nearly in tears over the tattered trees around her house, her windows also green with leaves.

I didn't see the worst until I walked this afternoon along the lake, where the storm's toll lay in a wide swath at the water's edge. On my beach, under a bright blue sky and beside incongruously calm water, lay several huge dead salmon, a dead Yellow-rumped Warbler, a dead House Sparrow and a wing from some small bird. Sprinkled among the bodies were slim, yellow-and-black feathers, plant and wood debris—and hundreds of brown-and-orange Red Admirals and Painted Ladies. Many of the butter-flies were still alive. Some farther up the beach on dry sand flew off as I approached.

I began freeing butterflies that struggled under heavy, wet sand, setting them on the hot beach farther from the water. They dried off quickly, some flying within minutes. For half an hour I moved among them. Butterflies lit on my hands, my shoulders, my sleeves. I could feel their tiny legs in my hair, their cold-blooded bodies looking for warmth. Into my grief for so many deaths came these delicate creatures, which had been pounded all night by crashing waves and were now drying and flying away.

What were millions of Red Admirals and Painted Ladies doing out over the lake last night?

ॐ ॐ ॐ

MONARCHS AREN'T THE ONLY MIGRATING BUTTERFLIES

Every spring, huge numbers of Painted Ladies fly from the Southeast, repopulating most of the North American continent by the end of May. They are often accompanied by Red Admirals, a darker version of the brown-and-orange coloring. All summer the Painted Ladies reproduce, in several generations, the last of which hibernates through the winter as adults. Oddly, although Red Admirals are thought to migrate south in fall sometimes, Painted Ladies make no return trip.

FINDING PAINTED LADIES

The Painted Lady's favorite food, the thistle, is one of the most widespread plants in the world, which possibly accounts for the Painted Lady's status as one of the most widespread butterflies in the world. A thistle's a good place to find the medium-sized (smaller than a Monarch, bigger than a Cabbage Butterfly) brown-and-orange butterfly with "angled"—jagged-edged—wings. The inside wings are orange flecked with black, with black-and-white upper wing tips. Outside, marbled wings camouflage the Painted Lady on tree bark.

False Solomon's Seal

The leaves along the lakeshore still hang limp and ragged after the storm, but the woods seem unaffected. Especially contributing to the lush look is a plant with long, boat-shaped leaves and clusters of white flowers, and another that looks almost, but not quite, like it. I managed to identify some as False Solomon's Seal, others as Starry False Solomon's Seal. The alternating, big, smooth leaves along arching stems about a foot or more long stand out amid the lacy foliage of Dutchman's Breeches and trillium that covers so much of the forest floor now. The scars left where these leaves break off the rootstock are said to resemble King Solomon's royal seal.

There are quite a few species of Solomon's Seals. Some, I have read, are "true" and some are "false." I object to this label of "false," a label which is applied to many plant species and seems to me to insult its essence. If I can find an alternate name for a plant with an unflattering name, I use it. In this case, however, the words are used to divide Solomon's seals into two groups.

❧ ❧ ❧

SOLOMON'S SEALS, "TRUE" AND "FALSE"

Now is a good time to identify the different Solomon's Seals. The long, pointed, oval leaves along a single, arching, unbranched stem are so similar that you probably need the flower to tell them apart. Solomon's Seals are "true"—species with white or greenish bells hanging along the stem—or "false"—species with a cluster of white flowers at the stem tip. Here's how to tell the two most easily confused "false" types:

"FALSE"-TYPE SOLOMON'S SEAL	FLOWER
False Solomon's Seal	Plumed, branched cluster of tiny white flowers at end of stem
Star-Flowered Solomon's Seal	Single row of larger (1/4-inch) white flowers at end of stem

MAY 15

Chokecherry

How does one begin to deal with the explosion of flowering shrubs and trees in May? Clouds of pink and white are everywhere. I decided to start in my front yard, with a rangy, slender shrub that was covered with dangling, elongated clusters of frilly, white flowers. All my choices should be so lucky. The only blossoms similar to that are found on a much larger tree. In minutes I identified my shrub as a Chokecherry.

My guides can't seem to decide whether Chokecherry is a big shrub or a small tree; I found it in both guides. As with many other cherry trees and shrubs, the scraped bark smells a little like cherry cough syrup, but the shape of the chokecherry's flower clusters sets it apart from all the other big shrubby bloomers.

ᏊᏊ ᏊᏊ ᏊᏊ

CHOKECHERRY: EASY TO FIND IN SPRING

A tall shrub or a small, slender tree with elongated 4-to-6-inch clusters of tiny, five-petaled white blossoms is almost sure to be a Chokecherry. You won't need to get close enough to count petals, though. Only the (50-to-60-foot) Black Cherry has similar flower clusters. When in doubt, check the underside of a leaf: The main rib on a Black Cherry will be fuzzy. Chokecherry flowers from late spring to early summer throughout the northern states, Canada and in higher elevations in the West.

TAKE A FEW LEAF-EDGE WALKS

Just for fun, take a leaf-edge walk: Instead of trying to identify species, look for certain kinds of leaf edges. You might start by looking for singly-serrated leaf edges. The oval, pointed Chokecherry leaves are a good example: Each "sawtooth" is the same size. Chokecherry leaves are about as finely serrated as leaves can get; you'll find larger and coarser examples, too. On a second walk, look for doubly-serrated leaf edges: These leaves have at least two sizes of teeth. For an armchair run, check the illustrations in this book.

WHAT IS A SHRUB?

This is not as silly a question as it may sound. Shrubs can be confusing: Some shrubs grow into small trees and are included in tree guides; others that grow quite low to the ground may be found in wildflower guides. Basically, however, a shrub is bushy— a perennial woody plant that usually has more than one main stem and is smaller than a tree and larger than a wildflower.

MAY 17

Honeysuckle

Having gotten a running start with Chokecherry, I decided to try another bush in my yard. I knew, in a general sort of way, what this one was: Who hasn't sucked the nectar out of honeysuckle blossoms? The bush in my yard was about four feet high, with small, pointed, oval leaves and lots of oddly shaped flowers about as big as the end of a finger. Some of the flowers were white, some pale yellow.

I broke a branch end off, took it inside to identify and, to my surprise, easily confirmed my guess. What kind of honeysuckle was another story. I guessed mine was Amur Honeysuckle, because the twigs were hollow and the flowers were both yellow and white. There are all kinds of honeysuckles, however, both native and introduced, with yellow, pink and/or white blossoms, and telling them apart isn't easy. Still, I'll always know a honeysuckle bush when I see one. Honeysuckle has a way of growing that makes it easy to identify any time of year.

Later I went for a honeysuckle walk and found four different kinds of honeysuckle shrubs, and a honeysuckle vine as well.

🐸 🐸 🐸

HONEYSUCKLE: THE PAIR BUSH

It isn't the irregular, five-petaled flowers that makes honeysuckle so easy to spot, although they help. It's the way everything on a honeysuckle shrub grows in pairs. The twigs grow in pairs and so do the leaves, flowers and berries. Often the pairs reverse direction, a pair of leaves growing north-south, while the next pair sprouts east-west, and so on up the branch, making a dense, tangled hiding place for birds.

SIAMESE-TWIN VINES

Although honeysuckle vines may not flower in distinct pairs as honeysuckle shrubs do, on many species the leaf pair at the end of each new growth is fused together like Siamese twins. It's eye-catching and easy to spot. Bright, trumpet-shaped flowers often cluster above the fused leaves.

MAY 18

Scarlet Tanager

This has happened before: My right brain catches something odd; a color, a shape, a sound. My left brain immediately counters: "Aw, it's just a gull/rock/robin." Used to be, I'd heed that cynical message. I never wanted to be caught the fool, even when no one was around to ridicule me. I'm learning, though, to have faith in my hunches, as I did today when I glanced out the window from my computer and my eye caught a brief flash of red. "Aw, it's just a cardinal," ridiculed the left brain. I went back to my screen, then changed my mind and went for the binoculars. It wasn't a cardinal. It was a Scarlet Tanager.

I'm guessing that anyone who's seen a male Scarlet Tanager in full mating plumage—lipstick-red head and body, gleaming ebony wings and tail—will probably remember the time and place forever. It stops the heart. To find such fierce beauty in my own backyard was to experience awe. Scarlet Tanagers are not especially uncommon, but they are secretive and seldom seen, hunting down caterpillars in summer foliage.

If I hadn't followed my hunch, I'd have missed a real show: For over an hour, two adult male Scarlet Tanagers flitted among the leafing trees between the house and the lake, while I watched them from the deck.

Sometimes they perched quite close, sitting for long periods. Spring is definitely the best time to see a Scarlet Tanager. By late summer, the male's bright feathers become patchy with green. During fall and winter, he looks much like the female: a greenish yellow, with black wings and tail.

ɞ ɞ ɞ

THE BLACK-WINGED RED BIRD

You can't miss a male Scarlet Tanager in mating red: It's our only red-bodied bird with solid black wings—no white on them at all. The red is scarlet, not the ruby-red of cardinals. It has no crest. The only bird resembling the Scarlet Tanager is the Summer Tanager, but its wings are red as well as its body. The females of both species are yellow-green and hard to tell apart. See them in the eastern half of the United States.

MAY 19

Great Blue Heron

This morning I was startled by a huge bird that floated across the road in front of me. It landed on the edge of the dune, in Mary's shrubs. Against the iron green lake beyond, I saw the white head of a Great Blue Heron, its showy black plume shooting behind it. I walked not twenty feet from it, but the heron seemed to think that if its body were hidden, then its head would be, too.

Great Blue Herons are wary birds; never before have I gotten anywhere close to one. I see them wading along the far edge of the lagoon and sometimes I see one flying along the lakeshore with a strong, slow flap of enormous wings, long neck kinked back. I'd never seen one along the exposed, sandy beaches.

Recently a friend told me he'd watched a Great Blue Heron toss a muskrat into the air and try to catch it going down headfirst. It took several tries. Getting airborne was awkward afterward, thanks to the grotesque bulge in its neck. Herons are known for their ferocious pursuit of food. That long, yellow bill will snap up almost anything: fish, frogs, even mammals.

⁊⁊ ⁊⁊ ⁊⁊

THE BIGGEST HERON OF THEM ALL

A tall, blue-gray wading bird with a long, black plume off the back of its head is a Great Blue Heron. Over a yard tall, the Great Blue Heron is our biggest heron, and that includes the egrets, which are actually herons, too. Our only other tall, gray wading bird is the plumeless Sandhill Crane. The only other dark, plumed heron is the Louisiana Heron, common along the Atlantic and Gulf coasts. It's a foot smaller and its plume is white.

MAY 21

Baltimore Oriole

Early this afternoon I was basking with Ellen in a spot of sun on her studio deck when a loud, liquid, seven-note whistle poured out of the canopy of leafing branches. "The Baltimore Oriole's back!" exclaimed Ellen, and sure enough, a big black-and-orange bird emerged and lit on a fresh orange half nailed to a nearby tree. Before long, a second bird, just like it, appeared, and the two brilliant males chased each other around the yard in a none too friendly manner for almost half an hour. In all this time, no yellow-breasted, brown-topped females were to be seen.

I expect we were treated to a springtime boundary squabble. Like Red-winged Blackbirds, Baltimore Oriole males arrive before the females to establish their territories. This was a prime nesting place: Not only did Ellen regularly provide fresh oranges, one of their favorite foods, but an oriole nectar feeder hung nearby in the neighbor's backyard. Baltimore Orioles often return to a favored nesting area.

I assumed that writing about a bird as bright, bold and brassy as a Baltimore Oriole would be fairly straightforward, but the Baltimore Oriole is neither an oriole nor from Baltimore. Resembling European Orioles, it was misnamed by immigrants: .Lord Baltimore was honored because his personal colors, black and orange, resembled the bird's. Further muddling matters, scientists, having discovered that the eastern Baltimore Oriole interbreeds with the similar-but-not-quite-alike western Bullock's Oriole where their ranges overlap, recently renamed both birds "Northern Orioles." I don't see much point in this renaming; no one I know has ever heard of the Northern Orioles (which happen to be common in the South as well) but they have loved the Baltimore Oriole for years. Not one to favor logic over love, I personally will continue with "Baltimore."

❧ ❧ ❧

FIREBIRD

A large (7-to-8½-inch), black-hooded, orange-bodied bird is likely a Baltimore Oriole, also called the Firebird. It has black wings and a black-and-orange tail. There are other black and orange birds, but they lack the black hood. The smaller, similar Orchard Oriole is hooded, but its body is chestnut, not orange.

MAY 23

Sassafras

Now that the leaves are coming out, I can begin naming the trees in my yard, and the easiest seems to be a group of trees—some quite tall, others slim, smaller versions—that are beginning to screen my view of the lake. Thanks to the short ones, I can reach the leaves, which come in wonderful, smooth-edged shapes: egg, goosefoot and mitten. I have found a wonderful book for identifying trees—*The Tree Identification Book*, by George W. D. Symonds (see Bibliography)—which is too large to carry on treks, but since it's a simple matter to bring a leaf home (if I can reach one), the bulk of the book doesn't bother me. I found in its pages, painlessly, the odd-shaped leaves: They were Sassafras leaves, an identification confirmed when I sniffed root beer from a crushed twig.

Found throughout the eastern third of the United States, Sassafras leaves turn buttery yellow in autumn, falling in a thick, glowing carpet of soft puzzle shapes.

ᥬ ᥬ ᥬ

TWO TREES WITH THREE LEAF SHAPES

Call a tree a Sassafras if it has almost hand-sized, smooth-edged leaves in three shapes—egg (no lobes), mitten (two lobes) and goosefoot (three lobes)—especially if the twigs smell like root beer. A Red or White Mulberry has similarly shaped leaves, but they are smaller and the edges are toothed. Sassafras berries resemble tiny, red-capped blueberries; Red Mulberries are more like blackberries; White Mulberry fruits are white.

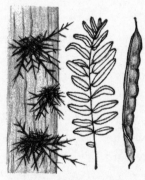

MAY 24

Honey Locust

Still trying to name the deciduous trees in my yard, I picked a large tree by the driveway that was frilled with many tiny, smooth-edged leaves paired along long stems. Each of these stems turned out to be a leaf, a compound leaf, made up of many smaller leaflets. Compound leaves, I've discovered, are a handy feature for identifying trees, because most tree groups have single leaves. Although it's not really possible to tell by leaf size whether a tree's leaves are compound, there is a simple way.

Trees with many pairs of small leaflets usually hail from the very large legume family, nearly all with very pea- or beanlike leaves and pods. Despite the many similarities, however, my tree was easily identified.

HOW TO TELL A LEAF FROM A LEAFLET

A *leaf* can be a single leaf or a group of several to many *leaflets*, making it a *compound leaf*. Every leaf on a deciduous tree has a little bud at the branch. If a leaf hasn't got a leaf bud, it's a leaflet, part of a compound leaf.

❦ ❦ ❦

A PRICKLY CLIMB

A tree with paired ranks of small, fairly smooth-edged leaflets (eighteen or more per leaf) is probably a Honey Locust if it has sharp, spiky thorns on the trunk and branches. Look also for twisted, inch-wide pods that resemble giant beans—up to a foot-and-a-half long—which sometimes remain on the tree all winter. Exceptions: Thornless, podless Honey Locusts do exist, available from commercial nurseries. A southern version called the Swamp Locust has the thorns but smaller, rounded pods.

MAY 25

White Ash

I thought that the tall, straight, slim tree that tops out over the second-floor deck, providing a convenient perch for chickadees to hammer open sunflower seeds, was another Sassafras until I noticed that what I'd taken for mid-sized leaves were actually leaflets—seven to a stem. It was surprisingly easy to identify: A tree with compound leaves, it turns out, is usually an ash, a Mountain Ash (which is not really an ash at all), some sort of nut tree or a legume.

My tree was an ash, but twelve species of ash were described in my tree guide. I narrowed these to the four common to the northeast quarter of the country. Mine was White Ash, the only ash with (usually) smooth leaflet edges and fairly large (2½- to 5-inch) leaflets. Nearly all the leaves contained seven leaflets each, typical of White Ash, probably the most common ash in eastern America. Suddenly I am seeing White Ash everywhere, including one great clump rising tall and straight as columns from a ravine between nearby wooded dunes.

❧ ❧ ❧

BY THEIR FRUITS SHALL YOU KNOW THEM

Trees with compound leaves in straight rows (not fanned out as on a Horsechestnut) often have unusual and conspicuous seeds. Identifying a species isn't always easy, but if you can see the seeds, you can often tell its family:

FAMILY	SEED CONTAINERS
Most nut trees (e.g. Walnut, Pecan, Hickory)	Nut, in fleshy husk
Legume (e.g. Locusts, Mesquites)	Bean- or pealike pod
Mountain Ash	Red berry clusters
Ash (e.g. White Ash, Green Ash)	Clusters of papery paddle shapes

MAY 26

Northern Whitecedar

At six o'clock this morning, I headed my white van north out of Douglas, and four and a half hours later, I pulled into Charlevoix, a popular lakeside town not far from the tip of the Michigan mitten (see map on page 135)*. By noon, we were pulling away from the dock on the Bea-

---

* Author's aside: As I mentioned earlier, I had to vacate my lakeshore apartment in the summers. To this end, I had rented, sight unseen, a one-room cabin on Beaver Island, a remote, sparsely populated island located at the top of Lake Michigan, a two-and-a-half-hour ferry ride from Charlevoix. I chose Beaver Island because it had long been a dream of mine to live in a cabin on an island, and I was able to find one there that I could afford. Research reassured me that the wildlife and plant species on Beaver Island were abundant

ver Island Ferry: my van below, hurriedly packed for the summer; myself above, at the rail, holding out my arms to the receding mainland. Soon fog wrapped the boat, and I fell asleep inside, to wake as we pulled up to the Beaver Island dock and the little harbor town of St. James. The trip had taken two hours.

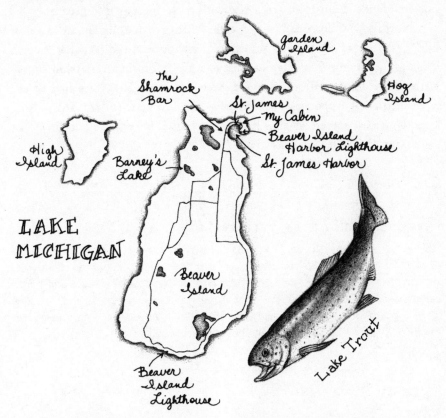

The harbor was located at the northern end of the island, protected by a "hand" of land that cups the northern edge, tipped by a lighthouse. Minutes after I'd received a hug from the harbor master, whom I'd met

and fairly representative of the mainland at that latitude, and Central Michigan University even ran an extension school on the island, mainly teaching courses concerned with nature and art. Although I'd worried a little that I'd be lonely on an island—writer or not, I didn't think I'd make a good hermit—that isn't what happened. The summer was joyfully social, as rich with new friends as flora and fauna. I grew so content there that I left the island only once all summer and, ignoring my van for my feet or bicycle, got by on a tank and a half of gas.

last summer, I pulled up to my new home, about a block from the lighthouse, a twelve-by-six-foot one-room log cabin, one of four nestled in the woods. Puzzle pieces of lake were visible through junipers, oaks and birches. At the back of the cabin, brushing against the window above the kitchen sink, was a large Northern Whitecedar. This island, as well as the surrounding smaller islands, I am told, are dense with these fragrant evergreens.

Northern Whitecedar is also called Arbor Vitae, meaning "Tree of Life," a name which several guidebooks attribute to the fact that its rich Vitamin C content saved James Cartier's group of explorers from scurvy in the early 1500s. I think it likely, however, that Cartier learned of the tree's medicinal properties—and name—from the native inhabitants, whose legends held it sacred, calling it the "Tree of Life," the "First Tree," or sometimes the "Tree of Knowledge."

᠀   ᠀   ᠀

THREE SCALY NORTHEASTERN TREES

Scaly evergreens native to the Northeast are easy to identify, unless you live on the Atlantic coast. Northern Whitecedar foliage is scaled and flat, with small cones (half-inch or smaller) and ranges throughout the northern northeastern states and northeastern Canada. Eastern Redcedar, which is really a juniper, has bristly ends and blue "cones," which look like berries (see page 35). Atlantic Whitecedar foliage resembles Northern Whitecedar, but is finer and not conspicuously flat.

MAY 27

Wild Columbine

As I was drinking my coffee out on my little front stoop this morning, I watched two hummingbirds hover below the drooping, red-and-yellow columbines that bloom profusely around my cabin and alongside the grassy two-track driveway that winds down to the road and the bay. The tiny birds flitted from columbine to columbine, ignoring the large patches of white-flowered Starry Solomon's Seal, but showing an occasional interest in my syrupy coffee. It was columbine they loved best, though. Their little heads tipped back as they shifted slightly forward to insert their long bills into the flowers, backed up to remove them, moved on.

A columbine blossom looks to me like the end of an ornate, goose-necked lamp: The five red petals form the shade, with the delicate spurs meeting at the top; the yellow inside is the bulb, and the dangling golden pistons and stamens form the glowing light.

Later in the day, my neighbor, Mary, who runs an incredible little toy museum and store next door, gave me a tour of her magical, voluminous garden, in which thirty-two kinds of lilacs were bursting into bloom. When I left Douglas yesterday, the lilacs were almost gone, which must mean that Beaver Island is almost two weeks behind southern Michigan. This seasonal lag is a good thing to remember; we are quite far north here. The days tend to be cooler in spring and summer, and warmer in fall, than in more southerly, mainland locations.

ಜ ಜ ಜ

FOUR KINDS OF COLUMBINE

All our columbines are variations on the same unique shape, making them easy to identify. Wild Columbine (red and yellow) is our only eastern columbine, flowering in the spring. Western columbines include Blue Columbine (blue and white), Longspur Columbine (yellow) and Yellow Columbine, which flower all summer, with Longspur Columbine holding out into the fall.

MAY 29

Cedar Waxwing

I have one of the best views in town. Across the harbor from the road, the little town of St. James, complete with picturesque church steeple and ferry at the dock, glows in the evening light. I've taken to sunset strolls along the road that curls toward town around the bay, but last night, I was lured up a side street by a lot of hissing and squeaking going on in a large-leafed tree.

I'd discovered a treeful of Cedar Waxwings. I was astonished at the sleek, dramatic beauty of the cardinal-sized, black-masked, crested birds. Like Wood Ducks, Cedar Waxwings look like objets d'art to me, too perfect to be real. Unlike Wood Ducks, however, Cedar Waxwings aren't at all shy. They act like good-natured flappers and dandies who know how to have a good time. Hanging out in crowds, Cedar Waxwings have been known to pass a cherry down a row of them, waiting for one to eat it, or gorge themselves on overripe fruit until falling-down drunk.

The Cedar Waxwing gets its name from the red, waxy dot on each of its wings and one of its favorite foods, Redcedar (which is really a juniper, not a cedar) berries. These birds love fruit, but will happily eat insects as well.

❧ ❧ ❧

SPOTTING A CEDAR WAXWING

If you see a crested, yellow-breasted bird, watch for a tail that looks as if the blunt end has been dipped in yellow paint. No other bird has that bright yellow band except the Bohemian Waxwing. Southerners need not worry about the difference, but the Bohemians share the northern half of the Cedar Waxwings' range (most of the United States and southern Canada). Although slightly larger and lacking the yellow breast, the Bohemian is almost the Cedar Waxwing's twin. In both species, male and female look alike.

MAY 31

Common Loon

Every morning I have been waking up to the unearthly laughter of loons, and some wild place in my heart laughs back. Whenever I hear them, I rush out to the shore with binoculars, but I never see one.

Today, on a tour of the island, I was sitting alone on a campground beach watching ducks, cormorants, mergansers, gulls and terns fly by, when suddenly, a large, black-and-white "duck" seemed to appear from nowhere, right in front of me, about fifty feet out. I knew immediately it was a loon: The black head, the daggerlike bill and that Fifth Avenue elegance gave it away. Who hasn't seen pictures of this gorgeous bird, with its zebra-striped neckband, perfectly checkered back and bright red eye? But what really knocked my socks off was its size: A loon is a really big bird. It's as big as a goose and almost a foot longer than a mallard.

Many environmentally conscious people are worried about loons being chased from their habitats and caught in fishing nets, so I have always thought of loons as fragile. But despite its threatened status, the loon is a powerful bird built like a Mack truck. Its bones are heavy and nearly solid, not air-filled like most bird bones. It can dive three hundred feet, fly sixty miles an hour and swim underwater for minutes at a time.

I watched the lone loon fish for twenty minutes. It would dive, stay under for sixty seconds (I counted), come up for five seconds, go down for sixty, up for five. Each time it vanished, it would reappear in some quite unexpected place. I was hoping to watch it take off—a loon launch is an awkward business, sometimes requiring a quarter-mile run across the water—but the loon was still fishing when I left.

❧ ❧ ❧

WHERE THE LOONS ARE

The Common Loon needs privacy and solitude to breed, spending summers in the far northern states and throughout Canada, in fresh water and inland lakes, especially around islands. It winters in salt water, along both East and West coasts. All four loon species are found in North America, mostly in Alaska and the far north of Canada. Three loons winter along the West Coast: the Common, the Arctic and the Red-throated Loon. Two, the Common and the Red-throated Loon, also winter along the East Coast.

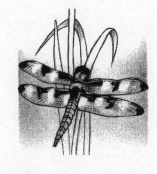

June

Little Blue Heron

 A lovely dark heron has been fishing with rigid concentration in front of Mary's house. I thought maybe it was a Sandhill Crane, because of its gray body and smooth, plumeless head. Don, my next door neighbor, also known as Wassak-waam, guessed it was a baby Great Blue Heron, because it was so slim and small. Mary set us both straight. "It's a Little Blue Heron," she told us. I looked it up in my bird guide, and it was.

I'd never heard of a Little Blue Heron, and when I checked the range map, I knew why: That Little Blue Heron had no business being way up here. According to the maps, it belonged in the southeastern, Gulf and Atlantic seaboard states. Some maps say it's "incidental" through the lower half of Michigan, but not this far north. Range maps, I am discovering, are at best a rough estimate of a species territory. The Little Blue Heron's color can be confusing, too: When immature, it's pure white, looking very much like an egret. Later it turns gray, but when an adult goes into mating season, its head and neck darken to purplish brown.

ð ð ð

DARK, LONG-NECKED WADING BIRDS

Of our three most widespread look-alike wading birds, only the Great Blue Heron (38 inches) is crested and has a yellow bill. The Sandhill Crane (37 inches) has a bright red crown. The smallest (22 inches), the Little Blue Heron, is crestless and crownless. If you live in the Gulf or Atlantic seaboard states, look also for the Reddish Egret, with its fringed, reddish head and neck, and the Louisiana Heron, which looks a little like a small, black-headed Great Blue Heron.

JUNE 3

Smooth Rose

Most wild roses, with their white or pink, five-petaled blossoms, oval, serrated leaves, and, of course, thorns, are hard to identify. Where their ranges overlap, they hybridize, puzzling nature namers. But on Beaver Island, and in much of the northeastern United States, there is a rose called *Rosa blanda*, the Smooth (or Meadow) Rose, which few rose enthusiasts make much fuss over, it being quite plain (which is what *blanda* means), if any rose can ever be plain.

Right now, this "plain" rose is turning Beaver Island into a pink-sprinkled fairyland. Not many of the two-inch blossoms adorn any one bush, but the low, upright shrubs thrive on dunes, along roadsides, around my cabin, along the shore, in meadows, in wet places and dry. So despite the confusion about wild rose I.D., I got lucky with this one: *Rosa blanda* is definitely different.

≈ ≈ ≈

THE THORNLESS ROSE

The Smooth Rose is the only really common wild rose that will never bite the hand that picks it. If it has thorns at all, they'll be down low, out of the way. The Smooth Rose blooms best early in June, but will bloom again later in the summer. Only a few pink, five-petaled, yellow-centered flowers bloom at one time on each bush, but, sprouting from underground runners, there are usually enough shrubs around to make a show.

JUNE 4

Snowshoe Hare

This morning two friends and I loaded a crate of four fat chickens on *Grandpa Johnson*, a small, gray fishing boat with an open back. Jon, the captain, Don and I took off to deliver our feathered passengers to a neighboring island, where they would spend the summer keeping campers in eggs. In an hour or so we arrived, put the chickens in a coop that also enclosed several cedar saplings and then gathered around a cement fire pit from which a stovepipe rose, piercing the huge blue tarp that covered the outdoor "kitchen." For several hours we sat sipping coffee with the residents, including Keewaydinoquay, a Native American botanist and medicine woman who had just recently arrived. We exchanged stories and watched small, white-patched "rabbits" hop past us, in and out of the woods.

They were Snowshoe Hares, Kee corrected me, also called Varying Hares because their white winter fur turns brown in summer, an awkward transition that amused her. "I called that one Boots when I first came"— she laughed a deep laugh, pointing to a hare that by now had only one white boot—"because it had white feet." Another had looked as if it were wearing jockey shorts.

I've never really known how to tell rabbits and hares apart, and no wonder: To tell a hare from a rabbit, say the experts, look at their newborn young; hares are born furred and seeing, while rabbits are born naked and blind. This may add to my knowledge of rabbits and hares, but it doesn't help me identify the long-eared brown creature eyeing me on a summer morning. The Snowshoe Hares looked like Eastern Cottontails to me: They are about the same size (two to four pounds) and brown with white tails, but the hares seem lankier, with very large feet. The easiest time to tell some hares—especially the Snowshoe Hare—from rabbits is in winter.

The Snowshoe Hare is found in almost all of Canada and in our northern states, especially around the Great Lakes and New England and in higher elevations like the Appalachians, the Rockies and the Sierras.

> **WHEN IS A RABBIT NOT A RABBIT?**
>
> When it's a Jackrabbit. Jackrabbits—including the Whitetail, Antelope and Blacktail Jackrabbits—are not rabbits, but hares.

◆ ◆ ◆

NO WHITE RABBITS!

Any white rabbitlike critter you see in North America is a hare; we have no white wild rabbits. The "Snowshoe Rabbit" is actually a Snowshoe Hare. Of the six common species of hare found in North America, three of them—the Snowshoe Hare, the Arctic Hare and the Whitetail Jackrabbit (a western hare)—turn white in winter, and one—the Antelope Jackrabbit (a southwestern hare)—has a white belly and sides all year.

SUMMER HARE HELP

Most hares have powerful back legs and huge ears. Size can also help distinguish a hare from a rabbit. Most hares are bigger than rabbits. The Snowshoe Hare is most difficult to distinguish because it is so small: 13 to 18 inches, about the size of an Eastern Cottontail, which is about as big as rabbits get. To tell an Eastern Cottontail from a Snowshoe Hare, look for white on the Cottontail's feet.

JUNE 5

Bracken Fern

The word *bracken* has a fairy-tale English forest sound to me, although I've never known what it meant. The truth is, a *bracken* is simply a place overwhelmed with enormous Bracken Ferns, our most common fern, found throughout North America and known all over the world. Beaver Island is practically a bracken unto itself: Bracken Ferns crowd the road-sides, around my cabin and the cedar woods so densely that little can grow beneath them. The Bracken Fern is one of the few ferns that doesn't require damp and shade; it thrives in dry and burned-over places.

The several varieties of Bracken Fern are all coarse, tough and tall. Some, especially in the West, form single, four-foot fronds. The one found here is three feet high or more and triple-branched.

🐦 🐦 🐦

BRACKEN: THE EASIEST FERN

An enormous fern growing out in the sunshine in dry, hard soil is likely a Bracken. A triple-branched Bracken is hard to mistake for anything else. Its yard-long fronds set it apart from other triple-branched ferns, like the smaller Rattlesnake Fern (10 inches) and the even more delicate and tiny Oak Fern (5 inches). Even the single-frond Brackens distinguish themselves by size, if not location.

JUNE 7

Poison Ivy

Each of the four friends who asked me what I was writing about today launched into a perfectly horrible story about a personal encounter with Poison Ivy. The adage each of them knew— "Leaflets three, let it be"—hadn't been enough. As I took a closer look at the lush crop bordering my cabin, I finally realized why. Though at a respectful distance from the Poison Ivy, I was standing barefoot in three-leafleted Wild Strawberries, while my knee brushed some kind of bramble that was also leafed out in threes.

It's no wonder I've never been able to sort out this Poison Ivy business. If I can't touch "leaflets three," I could well miss out on strawberry short-cake and blackberry pie, without which what is life? I find it easier to define Poison Ivy in terms of what it does *not* have (see below).

Poison Ivy is found just about everywhere in North America, but it's

---

"BERRIES WHITE, TAKE FLIGHT"

Poison Ivy, Poison Oak and Poison Sumac are related, all sumacs, flowering from the stem in clusters of greenish or white flowers, which develop into white or greenish berries.

Poison Oak leaflets grow in threes, but it is usually a shrub. The leaflets often resemble oak leaves.

Poison Sumac's seven to thirteen leaflets are smooth-edged, unlike the serrated leaflets on most nonpoisonous sumacs. The twigs are never fuzzy.

---

not the only rash-inducing plant. Poison Sumac is found in the East, while Poison Oak is common in the West Coast and southern states.

Don't take a chance with these highly toxic plants. Reactions can be long and painful. Although one friend claims to have cured a full-body outbreak in one day by sitting for six hours in the ocean, it's the only quick fix I've ever heard of.

❧　❧　❧

POISON IVY: NO THORNS, NO SAW-TOOTHED LEAVES

Many three-leafleted plants are rose or berry bushes. Prickly stems and/or sharply serrated leaf edges are not likely to give you a rash. But don't touch any part of any plant with smooth-edged, triple leaflets on smooth stems.* The leaves can vary in shape; sometimes they're wavy or widely toothed. The warning applies to almost any plant. Poison Ivy can appear as a wildflower, vine, shrub or small tree.

* A few "safe" plants also fit this description: the Hop tree and Prairie Rose, for two.

JUNE 8

Eastern Kingbird

This morning a good-sized flycatcher perched on a telephone line didn't even flick a feather as I walked by. I knew even without binoculars that that brazen bird was an Eastern Kingbird. The kingbird isn't called "king" for nothing! It'll chase off much bigger birds, even crows and hawks, if they dare to encroach on its territory. I've been seeing kingbirds everywhere, perched on telephone lines, fence posts, usually quite prominent places, and never budging an inch when I stop my car to point my binoculars at them. But it wasn't until I noticed the white band at the end of their tails that I knew for sure what they were.

🐦 🐦 🐦

KINGBIRD: THE EASIEST FLYCATCHER

The Eastern Kingbird is the only songbird with a white band at the end of its broad tail. It should be no problem finding one: Not only is it big for a flycatcher, but the Eastern Kingbird often chooses look-at-me perches. Its parts are in higher contrast than most of the other flycatchers, too: a white breast with dark gray wings, black head and (white-tipped) tail. Except for the West Coast and most southwestern states, the Eastern Kingbird is common throughout the United States and Canada.

JUNE 10

Yellow Warbler

I've been hearing wonderful warblers on my early morning strolls to the lighthouse. When I first arrived, the dominant theme was an insistent, long, loud *Sque-e-e-a-k*. The trees squeaked all day long, as if their joints needed a good oiling, but no amount of staring at the masses of leaves revealed the singers. After a few days, the noise ceased, the birds probably having departed for Canada.

This morning, I heard a new song: a cheerful, oft-repeated *Dee-dee-dee-diddly-dee! Dee-dee-dee-diddly-dee!* I searched the dense, dark recesses of the White Cedar from whence it came for ten minutes before a yellow bird flew out, poised like a golden ornament on the topmost spike, and resumed singing passionately. It was a Yellow Warbler. True, most warblers are yellow, but this one is *really* yellow. It's a great beginner's warbler; not only is it common throughout the United States and Canada, but it's a cinch to identify.

ə ə ə

ONLY ONE RED-STREAKED WARBLER

If you see a little yellow bird that is dull yellow on top and bright yellow on the bottom, look for red streaks on its bright yellow breast. If you find them, you've just identified a Yellow Warbler. It's also the only yellow-breasted warbler with yellow on its tail. It's not as shy as some warblers and is often happy to sing for you. *Dee-dee-dee-diddly-dee!*

JUNE 11

Northern Lights

Last night at about midnight, I was awakened by a knock on the door. "Sorry to wake you up," apologized Jon, "but we thought you'd like to see the northern lights!" Never having seen the northern lights, I pulled on my gray sweatsuit and went out into the chilly dark to join Jon, Mary and Glen in a small open space next door. Straight above us flashed a huge, fuzzy, white "cloud" that radiated moving beams of light; it could have been a spaceship hovering over the island in a protective shroud. Long smoky fingers of light streaked from the horizon to the center and back down again, jet trails, comet tails, darting across the sky. At the northern horizon, light bubbled, flared and glowed like a city with a dozen new service stations. We watched for about twenty minutes, heads thrown back, necks cricking, before succumbing to mosquitoes.

I couldn't sleep. At 2 A.M., I went out again and lay on my back on Jon's dock. This time, the energy centers had concentrated in two places, at opposite sides of the sky, and were shooting long white beacons toward each other. Bright white beams streaked from one side of the horizon to the other and then back, in parallel bars, or they came from both sides at once and met in the middle. Light danced over the entire sky. An owl hooted in the nearby trees. Now and then I'd hear a loud splash. Ducks quacked sleepily under the dock. Mosquitoes whined. A light breeze shuffled through the trees. It was a long time before I went to bed.

The next morning, I consulted by telephone an astronomer friend at the Jesse Besser Planetarium in Alpena, Michigan. Jim explained that the appearance of the northern lights had occurred because a solar flare* about six days ago had caused a strong solar wind. Particles (electrons and protons) were shot into the earth's atmosphere. "Generally, the wind is not so strong, and most of these particles are deflected around the earth, like water around the prow of a boat," he said. "But a high solar wind can penetrate the earth's atmosphere." Particles as dense as 100 million per square inch collide with atmospheric gas molecules fifty to five hundred miles above the surface, giving off light far brighter than all the cities of North America combined. The colors, sometimes intense, depend on the gases encountered and at what altitude. Generally, oxygen causes red at one altitude and green at a higher one, and nitrogen can occasionally cause turquoise.

The northern lights do not happen immediately. The particles collect in the earth's *magnetosphere* and are drawn toward the two magnetic poles. It can take a couple of days for the magnetosphere to become overloaded and the show to begin, but it can then go on for several days.

---

* A sudden, explosive release of energy on a small area of the sun

೩ ೩ ೩

WHERE AND WHEN TO SEE THE NORTHERN LIGHTS

The brightest northern lights, also known as the *aurora borealis*, usually fluoresce between fifty and a hundred miles above the earth and are seen with the most drama and frequency, whenever the sun becomes active, about a thousand miles south of the magnetic North Pole, but sometimes as far south as the Caribbean. Their frequency increases with the frequency of solar flares, which follow an eleven-year cycle. The last cycle peaked in the winter of 1990, but the years before and after have provided good shows, too. The next peak year will be 2001. (There are Southern Lights, too, called the *aurora australis*, visible in the southern hemisphere.)

JUNE 12

Hawkweed

This morning I took my coffee out to a rock by the harbor where a little wetland has formed, enjoyed by mallards, mergansers, gulls, terns, blackbirds, herons and occasionally a family of three Mute Swans. I noticed a new array of flowers blooming among the green grasses, reeds and willows, which I later identified as Ox-eye Daisies, Common Buttercups, Field Hawkweed, which resembles a gangly, teenage Dandelion, and its near twin, Orange Hawkweed.

It wasn't surprising that I'd found two kinds of hawkweed in one day: Hawkweeds are possibly the largest genus of flowering plants, with sixty-eight hundred species counted worldwide. (Although I couldn't find a figure for North America, at least twenty-five species are found in our western states.) Most of our hawkweeds have small (half-to-one-inch), rayed, Dandelionlike flowers, which can be yellow, orange or white. The plants fall into two groups: those with leafless stems and those that grow leaves on their flower stalks. Field and Orange Hawkweed fall into the first group.

Hawkweed gets its name from an old European legend which claimed that hawks ate it to improve their eyesight. Some species of hawkweed were imported from Europe as treatments for eye diseases. No helpful effect on vision or eye disease, however, has been proved by science.

⋙ ⋙ ⋙

TWO EASY HAWKWEEDS

If you see a cluster of little "dandelions" on a single, tall, thin, hairy stem growing from a small rosette of hairy leaves, you are probably looking at Field Hawkweed if it's yellow, or Orange Hawkweed if it's orange. There are similar yellow hawkweeds, but often the stems are smooth, only one flower tops each one, or leaves grow from them. Orange Hawkweed is easiest to spot, there being few truly orange flowers. (The western Orange Mountain Dandelion has toothy, Dandelion-type leaves.)

JUNE 14

Michigan Brown Spider

I came home from a bike ride this afternoon, reached out to open the screen door and almost grabbed a large spider sitting on the handle. I was pleased I'd seen it in time, since Mary tells me it can give a nasty bite. Its brown legs were not obviously striped, as many spider legs are here, and its head was very small. It was brown-bodied, with a beige pattern imprinted on its large, pebblelike abdomen. The body was about half an inch long, big enough to "pop" if stepped on. This method of spider control is frowned upon by Mary and some others here, who prefer to relocate their arachnids, still kicking, outdoors.

Out of the thirty thousand species of named spiders in the world (representing only one-quarter of the population), only sixty-three were represented in my insect and spider guide, mine not among them. Mary told me this spider, very common on Beaver Island (and in my cabin), was a Michigan Brown Spider, which is good enough for me.

The Michigan Brown isn't the only spider spinning webs daily in my cabin, behind the toilet, between my car antenna and the car door, across the clothespins on the laundry line, over the lamp shades and connecting the garbage can lid to the house. There are a number of tiny spiders, spinning tiny webs. I've seen slim, brown, medium-sized spiders, minute black spiders with white-spotted abdomens, spiders with pink lines and green spiders. None are pictured in my insect and spider guidebook.

         ❧ ❧ ❧

HOW TO IDENTIFY SPIDERS

It isn't easy. There doesn't seem to be a big interest in spiders, despite their fascinating behaviors—web spinning, jumping, hole digging—and their incredible variety. The best guide I've found is a small Golden Guide: *Spiders and Their Kin* by Herbert W. Levi and Lorna R. Levi (see Bibliography).

JUNE 15

Timothy

I've been captivated by the beautiful grasses that grow along the shoreline. There were so many kinds when I began to look: I saw tall, lacy, purplish stalks; fluffy, flowery green stalks; compact taillike grasses; delicate grasses with slim leaves. Today I collected eight kinds, some of which were already shoulder-high.

Up until now, all grasses have looked pretty much alike to me. They seem to be the spiders of the plant world: There are thousands of species, which appear to us daily, but we don't really see them and a comprehensive guide is not commonly available. There just doesn't seem to be a lot of interest in grasses. A few days ago, though, I found two small guides to Beaver Island plants that included sections on grasses. These have given me the courage to try to start sorting them out. Even with this help, however,

---

HOW TO TELL A GRASS FROM A SEDGE

The rule of thumb is this: Grasses are hollow, but sedges have edges. While this may not always be true, sedges tend to have triangular, solid stems, while grasses have hollow, segmented stems. Bamboo, for instance, is a grass. A common sedge is Porcupine Sedge (1 to 3 feet tall), with bristly, small (1 to 2½ inches) flowerheads.

I could identify only one of my grasses for sure: a calf- to knee-high grass with a compact, two- to three-inch spike that resembled a miniature green cattail. It is called Timothy and is often grown for hay. The grass blades seem to grow sparsely—three or four—from the same side of the stem.

Most people don't bother telling "grasses" apart—many grasses, I've discovered, require the use of a dissecting microscope to identify them positively—but it is satisfying to at least know a grass from a sedge, which resembles a grass and usually grows among grasses in wet places.

JUNE 16

Great Bulrush

Very early this morning, I saw the Little Blue Heron again, standing in what, until today, I have been calling "reeds." These are actually rushes, in particular the Great Bulrush. The shore here is dense with Great Bulrushes, and other rushes, too; Mary says four kinds thrive along her lake frontage. The long (up to six feet), round stem of the Great Bulrush here is topped by a small, woody flower cluster, over which curves a pointed, curved hook, like a tiny bird-head with a slender, long beak. Later in summer, the "head" will blossom into papery, dangling flowers.

The Great Bulrushes here harbor much wildlife: Red-winged Blackbirds nest in the ones close to shore; ducks and mergansers hide among

---

HOW TO TELL A RUSH FROM A GRASS

Although both grass and rush stems are round, grass stems grow in hollow segments, while rush stems are usually smooth and unsegmented, with pithy, white centers.

---

them farther out. They grow slim, green and graceful, and seem longer than they are, mirrored in the bright surface of the lake.

JUNE 17

Common Yellowthroat

Recognizing birdcalls has added a new sound track to my morning lighthouse walks. It's as if a song suddenly had words. Even if I can't actually see a bird tucked into foliage, I often can picture it there. This morning I picked out the yodeling and territorial racket of many blackbirds, the screams of jays, the *Wheat! Wheat! Wheat!* shriek of a sandpiper (see page 219), the twittering of darting Cliff Swallows, the melodious robin, the mellifluous Mourning Dove and even the Yellow Warbler's *Dee dee dee diddly dee!*

Then, as I sat on my favorite rock overlooking the harbor, a new song burst from a little tree right next to me: *Witchity ah witchity ah witchity ah witchity!* It was not a shy song. I stood around the tree for a while, peering with my binoculars while the song went on undisturbed—*Witchity ah witchity ah witchity ah witchity!*—before I saw it.

I never cease to be amazed at the jewellike birds that have lived over my head for nearly fifty years, while I walked beneath them, unaware. Today, I was once again stunned by the common but extraordinary beauty of my singer: Sporting a wide, black mask in startling contrast to the bright yellow breast below and the border of white above it, it was a Common Yellow-throat. Common, perhaps, but I had never seen one in my life.

🐦 🐦 🐦

LOOK FOR THE YELLOWTHROAT'S MASK

If I were to name the Common Yellowthroat, I'd call it the Lone Ranger for the wide, black mask that wraps right over its beak. The Cedar Waxwing and Magnolia Warbler also have masklike markings, but theirs are more like eye patches.

JUNE 18

Blueweed

Of the many flowers blooming now, one that stands out among all the yellow, white and orange is an intensely blue-violet flower in clusters of red-stamened bells up a tough-looking, hairy stem. When I first looked at the flower, I thought it was diseased; the bristly stem was polka-dotted with tiny dark specks. This, however, turns out to be the thing to know it by.

This flower is commonly called Viper's Bugloss, a name so ugly and so hard to say, much less remember, that I've used Blueweed, another of its common names. Kee told me during my last visit to her camp that, like myself, she objects to names given to plants that dishonor them, such as names that use the word *least, stinking* or *false*, or names that compare one plant unfavorably with another.

There are so many plants that bloom with elongated clusters of blue or violet flower spikes that they all look alike to me. A closer look, however, can solve some of the problems. Many blue and violet (or blue-violet) flower spikes are made up of oddly shaped blossoms, often variations on the snapdragon shape. Blueweed flowers, however, are *bell-shaped* blossoms.

ใช ใช ใช

BLUE-VIOLET, BELL-SHAPED WILDFLOWERS

The only flower you're likely to confuse with Blueweed is the Bellflower, whose stem is also hairy, but lacks the Blueweed's speckles. Here are some common bell-shaped, blue-violet wildflowers:

NAME	STEM	FLOWERS
Blueweed	Hairy, speckled	Borne in clusters up the stem
Tall Bellflower	Hairy, not speckled	Bloom along main stem
Virginia Bluebell	Smooth	Dangle in clusters
Harebells	Smooth	Dangle individually
Downy Gentian	Downy	Show five petals when open

JUNE 20

Cliff Swallow

I've often watched the many swallows here, but they darted over the water in the light, haphazard way of butterflies, too quick to catch in my binoculars. I thought, There's got to be some way to tell swallows apart, even in fast flight. The sleek but sparrow-sized birds have always looked alike to me. This morning I took my coffee and binoculars down to the lighthouse, where swallows swoop in and out of mud nests crowded under the white brick tower's black metal overhang.

Settling on a driftwood log next to the quiet water, I focused on a particularly active group of nests, watching swallows flutter in, hover, vanish into jug-shaped nests through little round holes, fly out again. Little by little, I began to *see* them: the black-topped wings, the orange rump, the chestnut-and-black head, the white patch above the beak, the white breast.

They were Cliff Swallows, famous for their punctual spring return (usually March 19) to California's Mission of San Juan Capistrano.

☙ ☙ ☙

TELL SOME SWALLOWS BY THEIR HOMES

Many swallows, which tend to nest in colonies, aren't as secretive about their nest locations as most other birds, and some swallow homes are unique. Cliff Swallows return often to jug-shaped mud nests built under eaves or in other high, protected places. Swallowtailed Barn Swallows swoop into empty buildings or barns, to cup-shaped mud nests. Brown Bank Swallows dart in and out of holes in mud or sand banks. The Purple Martin, the only all-dark swallow, often nests in white, many-holed martin houses perched on tall poles.

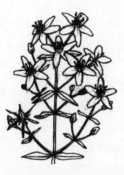

JUNE 22

St. John's-wort

Today is Summer Solstice, and the bright yellow blooms on the many two-foot-tall St. John's-worts are puffing out right on time. Because this sunny flower starts blooming right around the third week in June, it was long used in Solstice celebrations. John the Baptist's birthday is said to be around Solstice, too, so eventually, the plant picked up a Christian name. Bright and yellow, with a bushy spray of stamens bursting from the middle like a delicate package decoration, St. John's-wort flowers are a fitting gift to the sun on its biggest day.

St. John's-wort has a strong reputation around here as a cure-all. Some Native Americans dry the plants for herbal remedies, and I'm told that the yellow flowers turn purple when they're boiled. I have an awful time distinguishing the many five-petaled, yellow flowers that are blooming right now. They all seem about the same size and shape. Upon close study, however, I have found some ways.

ò ò ò

SOME FIVE-PETALED, YELLOW LOOK-ALIKES

Most St. John's-worts—there are several varieties—have puffy flower middles and smooth leaves. When in doubt, hold the leaves up to the light: You will see many tiny, translucent dots in the leaves. Here are just a few other penny-sized, yellow, five-petaled flowers:

YELLOW, FIVE-PETALED FLOWERS	FIELD MARKS
St. John's-wort	Bushy stamens, translucent dots in leaves
Buttercup	Glossy petals (five to seven)
Silverweed	Leafless flower stalks, silvery leaf undersides
Sulfur Cinquefoil	Hairy, five-fingered leaves

JUNE 23

Northern Water Snake

Last night I was watching the sunset from the dock when I noticed a protruding head, no bigger than the end of a broom handle, approaching through the bulrushes. Soon I could see the rest of it: a snake about a yard long, undulating gracefully through clear water. It appeared to be charging me and I wondered if I should flee. A few feet short of the dock, however, the snake dove about three feet to the bottom and swam directly under me, coming up for air on the other side of the dock. We eyed each other once more before it departed. I counted a full sixty seconds before I saw the head come up again, a black dot a good distance out.

I knew it was a Northern Water Snake because a few days ago I peeled a baby snake just like it off the road; it had been run over and was flat as paper. I brought it home and laid it next to its picture in my eastern reptile guide. I don't often get to make such a positive identification of a snake, since my glimpse of one is usually fleeting, and I'm shy about handling live

> ### CAUTION: A SNAKE IN THE WATER
> ### MAY NOT BE A WATER SNAKE!
>
> Although true water snakes are not poisonous, all snakes can swim, and many enjoy the water, including some that are poisonous. I've seen rattlesnakes in Rocky Mountain lakes, and the dangerous Cottonmouths (Southeast), which closely resemble several southern water snakes, are semiaquatic.

ones. Now I could look at the dead snake's belly, the pattern on which supposedly can be more helpful than the back in identifying snakes. I don't usually find this very helpful. Snakes just don't roll over for me.

🐌 🐌 🐌

NOTES ON THE NORTHERN WATER SNAKE

The Northern Water Snake is probably the biggest water snake in the Northeast (24 to 42 inches), best identified by the wide, rich brown bands that go around the top of the snake near the front end and become alternating blotches on most of the remaining length. (The colors can vary, depending on the area.) Sometimes called the Common Water Snake, it is common throughout its fairly wide range, which covers most of the northeastern quarter of the country, from eastern Colorado right into southern Maine.

JUNE 25

Eastern Milk Snake

Today on another visit to Kee I saw a snake tail sticking out from under the cement fire pit. I watched it a long time and it didn't move. I knew there were no poisonous snakes on the islands here, so I reached out and pinched it. My fingers closed on empty snakeskin: The snake was shedding, but when I tugged gently, the tail was definitely attached. I prudently let go. After a while, when the fire warmed up, a milk snake poked out its

---

EVEN NONPOISONOUS SNAKES WILL BITE!

It probably wasn't smart of me to pull on a snake's tail. Nonpoisonous snakes are "harmless," but many, including the milk snake and the Northern Water Snake, can inflict a painful bite if handled or harassed.

---

head, followed it with about ten inches of itself, looked us over for about a minute, reconsidered and this time fully withdrew. It was the prettiest snake I've seen yet: silvery gray with large and small, bright brown patches all over it, a little like a spotted cow.

The Eastern Milk Snake I.D. was made on the spot by someone with a long acquaintance with this resident, and when I got home, my reptile guidebook confirmed it, also informing me that the milk snake can grow to four feet. It was once believed that milk snakes milked cows, but, of course, they don't. They are constrictors which eat other snakes, but they also frequent barns and, along with cats and owls, make terrific mousers.

Not all milk snakes are spotted. Some milk snakes resemble king snakes, with bright rings, as well as the venomous Eastern Coral Snake (Southeast) and Western Coral Snake (Southwest). It also has been mistaken for the venomous Copperhead found in some middle eastern states.

🐸 🐸 🐸

JUST ONE NORTHEASTERN MILK SNAKE

Although there are quite a few species of milk snakes in North America, the Eastern Milk Snake is the only milk snake found in most of the Northeast. Eastern Milk Snake spots are in higher contrast, dark on light, but they do slightly resemble the Eastern Masassauga Rattlesnake's; the milk snake is slim, however, and lacks the enlarged rattlesnake head.

Eastern Garter Snake

Cindy, my slim, cheerful waitress at the Shamrock, informed this morning that I was writing about snakes, related this story: She had been repairing some sagging insulation in her new bedroom—she hadn't gotten the drywall up yet—when enough snakes to fill a bushel basket fell out! She stomped and yelled until the snakes slithered down the walls and disappeared. The next day, they were back. She chased them out again and still they returned. This problem is not uncommon here; it's not unheard of for balls of tangled, hibernating snakes to become so heavy that they fall through the ceiling. Remedies abound, including the playing of loud rock music; theoretically, although snakes are deaf, they are irritated by the vibrations.

These islands are full of garter snakes. Cindy's snakes were probably Eastern Garter Snakes, which winter intertwined in large numbers. I've seen six or seven snakes around my cabin that I took to be garter snakes, but I didn't know how to tell that for sure. As it turns out, it's fairly easy to spot a garter snake.

🐍 🐍 🐍

THE "STREAKED" SNAKE

A smallish (18- to 26-inch) snake with three light stripes that run the length of its body—one down the middle and one on each side—is likely a garter snake, named for the striped garters that used to hold up men's socks, although it can be a ribbon snake. Usually the stripes are yellow, but they can be other colors, too. The thirteen

---

HOW LONG IS A SNAKE'S TAIL?

You can tell the slimmer but similarly striped ribbon snake from a garter snake by its tail: The ribbon snake's tail is very long—one-third of its body length. But where does the tail begin? Everything after the anus is tail, say the naturalists. Unfortunately, this part of a snake's anatomy, being underneath, is not always easily glimpsed!

SNAKES WITH BODY-LENGTH STRIPES ARE NONPOISONOUS

No North American poisonous snake displays long, horizontal stripes. Most snakes streaked lengthwise with light stripes are harmless garter or ribbon snakes; the few other lengthwise-striped snakes are usually striped with darker colors (see page 165). Even these snakes are not poisonous, however, and most are limited to the Gulf and southern Atlantic states.

---

species of garter snakes in the United States include many subspecies, as well as pattern and color variations within species. A variation of the Eastern Garter Snake, for example, appears checkered and has no stripes at all.

JUNE 28

White Campion

A funny white flower—five petals spreading from a greenish, bladderlike sac—has been proliferating everywhere, but I put off writing about it, thinking there wouldn't be much to say. It's easy to assume that common things can't be very interesting. What I'm finding out, however, is that *widespread* and *common* are not always synonymous, as I discovered when I identified White Campion.

For starters, White Campion is a night-bloomer. The petals often twist up for most of the day, then spread out flat at night, when they are pollinated by moths. The flowers even smell more fragrant at night. Even stranger, White Campion produces both male and female flowers, borne on different plants. They're easy to tell apart: Male flowers bloom from elongated, tubular sacs, while sacs on the female flowers are puffed out and rounded. There are a number of white-blooming Campions and Campion look-alikes—including Bladder Campion, which is also common here—but it's easy to tell them apart.

ঝ ঝ ঝ

WHITE CAMPION AND LOOK-ALIKES

White Campion is the only one of all the white-petaled flowers with the typical campion sac that is *fuzzy* over most of its green parts. It grows on long (up to three feet) stalks with paired leaves and two or three flowers on a plant. Bladder Campion has smooth sacs and leaves and flowers in numerous smaller bunches. Starry Campion has fringed petals, and Night-flowering Catchfly is sticky. Most of these flowers can be found over almost all of the United States and Canada.

JUNE 29

Red-bellied Snake

Wouldn't you know, mere days after I declare that a snake with body-length lines is most likely a garter or ribbon snake, the next snake I see, left dead as a stick by my front steps, is an exception. Dark lines ran the length of its gray body, which was pencil-thin and pencil-sized, too, if you added a tail and rolled the whole thing thinner. I thought it was a baby, but it turned out to be, at nearly eleven inches, a rather large adult. It was easy to identify as a Red-bellied Snake, even without turning it over to expose the coral-red belly. The belly is not often the first thing one sees when a little snake zips by, and there are some other small snakes that are red underneath as well. Here's how to tell the Red-bellied Snake topside.

ঝ ঝ ঝ

A PIN-STRIPED SNAKE

Although the background color can vary, usually gray or brown, a tiny (8- to 10-inch), slender snake with four dark pinstripes is likely a Red-bellied Snake. If you can, try to get a look at its startling red (sometimes orange or yellow) belly, too. The Red-belly's range extends to more than the eastern half of the United States, including a narrow strip of southern Canada.

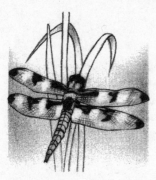

JUNE 30

Twelve-spot Skimmer

Today I drove about ten minutes to Barney's Lake and sat in the sunshine amid Giant Bulrushes and nine kinds of wildflowers to observe dragonflies. It was midday, which I'd read was an active time for dragonflies, and indeed, I counted at least ten kinds of dragonfly and one species of damselfly that darted and hovered around me. I tried to categorize what I saw into types, even if I couldn't identify the species, my insect guide including only thirty-six species. In an hour, I found examples from three of the five categories that were described in my Audubon Society *Guide to Insects and Spiders of North America*.

I didn't see any darners. Most of the dragonflies I saw today were skimmers, the largest group of dragonflies. I identified two. The most numerous were Twelve-spot Skimmers, with three black spots on each wing against a white (male) or transparent (female) background. They darted with fabulous speed and ease, even mating in midair, over the lake. Perched motionless on a cedar branch was a gorgeous, red-bodied Elisa Skimmer, with red blotches on the inner part of its black-tipped wings.

I thought I saw two kinds of Gomphids, but Gomphids are supposed to prefer streams, so maybe I was mistaken. I didn't see anything that seemed to fit the two remaining groups. I noticed huge differences in body shapes, though, from short, fat bodies to long, needle-thin bodies. The very thinnest bodies I saw belonged to the intensely turquoise-blue damselflies.

DAMSELFLY OR DRAGONFLY?

A damselfly rests with its wings together over a thin abdomen; a dragonfly rests with wings spread flat. Unlike dragonfly wings, both pair of damselfly wings are shaped the same.

ⅇ ⅇ ⅇ

FIVE KINDS OF DRAGONFLIES

I never did find a popular guide to the 450 species of dragon- and damselflies in North America, but I was able to place many of the specimens I saw into one of these groups:

Darners, enormous, brilliant green or blue, sometimes brown dragonflies with long tails, are what I think of when I think "dragonfly";

Common Skimmers' bodies are shorter than their wingspan and are often fairly fat;

Gomphids, unlike other species of dragonfly, have widely separated eyes. (The dragonfly's swiveling head is usually covered with eyes so huge that most meet in the middle, each eye containing up to 10,000 six-sided facets);

Biddies and Flying Adders are large and hairy;

Green-eyed Skimmers are skimmers with green eyes.

THE DRAMATIC DRAGONFLY

The dragonfly can hover as well as fly forward or backward on wings that, unlike those of any other insect, can move independently, each networked with veins containing as many as three thousand cells. It can fly up to thirty miles per hour and lift up to fifteen times its own weight. The dragonfly's life cycle, easily as amazing as the better-known butterfly's, includes a transformation from an ungainly aquatic nymph to an iridescent aerial acrobat.

July

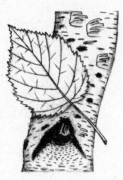

Paper Birch

On this beautiful, golden day, Mary and I went to pay our respects to one of the island's oldest trees, a spectacular Paper Birch with a girth larger than my arms could circle. Its thin, white bark peeled only in a few places on the massive trunk and on the many smooth, substantial branches that spread high and wide above us. The old tree presided over a sea of waist-high Bracken Ferns, the awesome base hidden from the road by shrubs and small trees. Two years ago, Mary and some other women celebrated the grand tree's four-hundredth birthday by leaving a cake wedged in its huge lower branches. They erected a sign nearby that announced the tree's longevity and requested that visitors refrain from declaring their love or presence on the sensitive bark. They have.

The Paper Birch is the only birch I can identify, thanks to its distinctive horizontally peeling white bark. But there's also Yellow Birch, Black Birch, Red Birch, Sweet Birch, River Birch and Water Birch, to name just a few. All of them have double-toothed leaves, but so do alders, elms, hophornbeams and hornbeams. Some birches are trees, but many grow among those look-alike shrubs that thrive along streams and other wet places. The best I can do as a beginner is sort out the few birches that adhere to some unique birch characteristics.

It's harder to distinguish birches in high elevations or northern climes, where more species proliferate. The three below are northeastern trees, although the Paper Birch is common in most of Canada and the states bordering Canada.

&a &a &a

THREE NORTHERN BIRCHES

All the birches have doubly-serrated leaves: leaves edged with different-sized teeth. Thin bark that peels horizontally, or twigs that smell of wintergreen oil will identify some birches.

Paper Birch:	White bark that peels in large pieces (also called the American White Birch)
Yellow Birch:	Silvery yellow bark that peels in shaggy curls; Wintergreen smell to twigs
Sweet Birch:	Nonpeeling bark; twigs smell of Wintergreen

JULY 2

American Toad

Walking home late last night, I was about to sidestep a blob that looked, by moonlight, like something left by the neighbor's Saint Bernard, when it moved. The turd was a toad. I yelled at Mary, who was just entering her house. She came over, glanced down and said casually, "Oh, that's Winston. Let's get him off the road." I gently nudged Winston toward Mary's yard with the side of my shoe. Winston, Mary informed me, had lived around the pond in her garden for some time.

Identifying toads can be difficult. Lucky for me, there is only one species of Toad on Beaver Island. Winston was an American Toad. Guidebooks describe the American Toad as being two to four and a half inches in length, but Winston was at least six inches long. For some reason, Beaver Island toads can become gigantic for the species. This is sometimes explained by age: Toads keep growing every year, and maybe some survive longer here.

I'm glad I didn't pick Winston up. Toads don't give people warts, but they'll urinate readily on handlers, and their excretions can irritate human mucous membranes; don't pick up a toad and then rub your eyes. Toads look so much alike that I'm not inclined to try to tell them apart, but I've always wanted to know how to tell one from a frog.

TOAD OR FROG?

I knew Winston was a toad right away, because, like many eastern toads, he had one light line down the middle of his back. Frogs with lines usually have two, usually yellow. Warty skin is another toad characteristic; frog skin is usually smooth. There are exceptions to both rules, though: Many toads, especially western toads, are lineless, and warty frogs and smooth toads are not unknown.

JULY 4 INDEPENDENCE DAY

Common Milkweed

Common Milkweed's starry flower clusters look to me like exploding pink fireworks, an apt subject for the Fourth of July. I've long recognized the large, paisley-shaped milkweed pods, because as a girl I used to collect them in the fall for their silk, to cushion my butterfly collection,* but I've never noticed the beautiful flowers before. Common Milkweed looks much like a small waiting-room rubber plant until it bursts into clusters of about seventy-five small, double-starred flowers each, up to six hundred flowers per plant. Rich in nectar, the Common Milkweed smells so sweet it rivals lilacs; a friend's claim that the French mix milkweed with rose petals to make perfume seems quite credible to me.

Milkweed flowers are not all sweetness and beauty. Insects lose their footing on the slick petals, ending up with a leg trapped in the flower's base. If the insect is strong enough to free itself, the leg will come out laden with a double dose of pollen, to be carried to the next flower, when the insect will slip again. If the insect cannot get free, it will die there. Common Milkweed also exudes a sticky, milky sap that is poisonous to most species.

* Butterfly collecting is no longer an approved hobby.

TWO THOUSAND SPECIES OF MILKWEED

Between seventeen hundred and two thousand species of milkweed thrive world-wide, many in North America. Although all North American milkweeds have the double-starred flower clusters unique to milkweed, they can differ in color. Common Milkweed can vary from light pink to dusky purple, but there are other rose-colored, as well as white, orange and yellow milkweeds.

Those that do eat it become poisonous, too, often painted in nature's danger colors: red (or orange) and black. This explains why Monarchs, which in caterpillar form gorge exclusively on milkweed, are rarely snapped up by birds and other predators.

🐢 🐢 🐢

HOW TO TELL COMMON MILKWEED

Common Milkweed often grows in clumps, springing up from deep, underground rhizomes. The only milkweed with warty pods, Common Milkweed is easiest to distinguish in the fall. In summer, look for flower colors varying from almost white to pink to lavender to dusky purple, even within the same group of plants. The leaf undersides feel deliciously downy. Common Milkweed is a frequent field and roadside plant, found throughout the United States, except in the far West.

JULY 5

Common Snapping Turtle

Yesterday morning, I awoke to gaze upon an enormous, live, gaping turtle at my door, dangling upside down, well away from the body of my friend Jon. "How do you like this one?" Jon said in greeting.

"Is it a Snapping Turtle?"

"You bet! A lot of meat on it, too!"

I became indignant, pleading for the turtle's life. Not only was this turtle twenty or thirty years old, judging from its size—the shell alone must have measured a foot and a half—but it was likely a female, since Jon had found it crossing the road. Female Snappers more readily leave the water than the males, trekking across dry land to dig a nest and lay their eggs— twenty to sixty, each almost an inch around—before returning to the water.

The Common Snapper is Michigan's biggest turtle and the only turtle in the state that can legally be captured and eaten. Who could possibly dine on a venerable mama turtle? Just about any islander, I discovered. Few here seem to suffer much remorse when a Snapper hits the frying pan. Snapping Turtles eat waterbird babies, I was told. They kill baby loons. Even Mary, an avid environmentalist, shared this view, but she felt for the turtles, too. "Ordinarily," she said, "Snappers wouldn't threaten loons, which prefer deep water. But in recent years, boaters have chased the loons into shallow water, and that's where the Snappers get them. It's not the Snappers' fault, really; it's ours."

A naturalist friend is concerned by the alarming decrease in the number of turtles in Michigan, so I have been braking for turtles and have scooped several off the highway. Each time I worried that I might be handling a Snapper. I'm relieved to know what a Snapping Turtle looks like and what to do if I need to pick one up (see below).

I am told by a turtle expert that the Snapping Turtle rarely bites a person underwater, but on land, its short temper is legendary. The Snapper is not really mean; it simply feels vulnerable because its shell does not close as effectively as most turtles'.

Fried in batter, our Snapper, which tasted like chicken, fed eight of us around a makeshift table outside my cabin. Snapper stories abounded. Mary told of shaping mud around Snapping Turtle eggs when she was a girl, and hardcooking them in a camp fire until the mud dried. Glen, the island's harbor master, once met a Snapper that bit a friend's backside and wouldn't let go. Glen himself had sawed the turtle's head off. My neighbor Don swears he's seen Snapping Turtles bite through broom-sticks, personally refuting all the articles I'd read that found such stories incredible.

I joined in the banquet reluctantly. Many naturalists frown on Snapping Turtle consumption, but is it any different than eating salmon or venison? It was a legal meal, but I won't do it again, siding, if belatedly, with the turtle.

≈ ≈ ≈

HOW TO TELL A COMMON SNAPPING TURTLE

Look at the tail. Not only is a Snapping Turtle's tail often as long as the shell, but it's saw-toothed on top, like a dragon's tail (except for the southern Alligator Snapper's tail, which is long, but smooth). Other freshwater turtle tails are short and smooth. To pick up a Snapper, grab the tail and hold the Snapper far away, with the underside toward you. Although all turtles are toothless, you don't want to be caught in the vise-like grip of a Snapping Turtle's beak—not even a small one.

THE BIGGEST FRESHWATER TURTLE

There are two species of Snapper in the United States: the Common Snapping Turtle (8 to 20 inches), found in most bodies of fresh water in the eastern two-thirds of the country, and the Alligator Snapping Turtle (13 to 26 inches), found in the southeastern states, especially along the Gulf. The Alligator Snapping Turtle can weigh as much as two hundred pounds!

JULY 7

Ox-eye Daisy

I have just learned an amazing thing: that those simple, white-fringed, yellow-middled daisies blooming everywhere right now are not flowers! They are *flower heads*, a collection of many flowers. Each "petal" is a complete, individual female flower called a *ray flower*. Over a hundred tiny, tubular *disk flowers*, each containing male and female parts, make up the yolklike middle. When the daisy is young, the disk flowers are closed up tight. Gradually, starting from the outside, the male parts are extended, pushed out by the female parts. As the Ox-eye Daisy ages, the middle gets fluffier. Most rayed "flowers"—asters, Black-eyed Susans and sunflowers, for example—are flower heads, too.

The Ox-eye Daisy isn't the only white-rayed, yellow-disked flower head: Mayweed, Chamomile and Easter Daisies look very similar. But there's an easy way to tell an Ox-eye Daisy.

᠀᠀ ᠀᠀ ᠀᠀

OX-EYE DAISY I·DENT·IFICATION

To tell an Ox-eye Daisy from daisy look-alikes, look for a dent in the middle of the yellow disk. All but the older Ox-eye Daisies are depressed in the center.

JULY 8

Queen Anne's Lace

I've been admiring the Queen Anne's Lace that's been frilling up the roadside out front, so I picked one and brought it home. It wasn't Queen Anne's Lace at all; it was Yarrow. This morning I walked through the cedar wood to Gull Harbor, and along the road there I found among the many blooming wildflowers another kind of lacy, flattish, white flower cluster.

This time I got it right: It was Queen Anne's Lace, each stem topped by a four- to six-inch cluster of smaller clusters of tiny, white, starlike flowers with one tiny purple flower in the center. Delicately pronged green *bracts* (modified leaves, usually at the base of a flower) collared not only the entire flower head, but each cluster inside it. The delicate effect was breathtaking. Women used to try to make lace as fine as this flower. When Queen Anne's Lace goes to seed, the flower heads close in the rain like little birds' nests, keeping the seeds dry.

All sorts of wildflowers blossom in flat clusters of small white flowers. I was lucky to get it right on just the second try. Queen Anne's Lace has fernlike, frilly leaves, so many look-alikes can be eliminated by their large or fairly simple leaves, but by no means does that take care of them all. If

---

QUEEN ANNE'S DEADLY RELATIVES

Socrates was not killed by hemlock tree tea but by the juice of the Poison Hemlock, a deadly plant that closely resembles Queen Anne's Lace in flower and foliage. Queen Anne's Lace's hairy stem immediately distinguishes it from its dangerous, smooth-stemmed relatives. Poison Hemlock's stem is smooth and spotted with purple. Fool's Parsley, another poisonous look-alike, also has a smooth stem. When crushed, the foliage on these two poisonous plants smells foul.

---

you know what to look for, though, you won't mistake any twin for Queen Anne's Lace.

🙠 🙠 🙠

COLLARING QUEEN ANNE'S LACE

A flattish, sometimes cupped, cluster of small white flowers with a green "collar" of delicate, three- to seven-pronged dragon's tongues flicking out beneath it is Queen Anne's Lace. Each smaller cluster within the large cluster is also ringed with little dragon's tongues. Known as "Wild Carrot," Queen Anne's Lace foliage looks and smells like carrot tops. Yarrow, with its white flower clusters and fernlike leaves, looks similar but smells quite different: strong and spicy. Often found together, both plants are common throughout North America.

JULY 9

White Clover

With beautiful wildflowers blooming faster than I can keep up with them, I thought I wouldn't bother with clover. But even the variety I thought was most boring, White Clover—the low-growing one with pinkish-white, inch-wide pom-poms, so often found in lawns—charmed me when I looked at it closely. Like certain violets, but unlike most clovers, each three-part White Clover leaf grows from the ground on its own stem, as does each bare-stemmed blossom. The flower head—a compact globe of dainty Sweet Pea–like flowers—begins blooming at the bottom. As each flower is pollinated, it closes up and droops, while new flowers bloom invitingly above. When I looked at each blossom of White Clover, I could tell just how far the pollination had progressed. A new flower head was tight and round, blooming only at the bottom; a half-pollinated flower head was skirted, like a ballerina; a dried flower head lacked only a face to resemble the head of a tiny, brown-haired doll.

ð›

THREE-LEAF CLOVERS

Clovers are called *Trifoleum*, meaning "three-leaved." If you're lucky, you might find a four-leafed clover, especially a Red Clover. Although all except specially bred clovers have three-part leaves, usually with small, oval leaflets, three leaves do not always a clover make. Some other pea family members, alfalfa, for example, have similar leaves.

JULY 10

White Sweet Clover

Clover is remarkable stuff. No one should live without it, and few have to: Most clovers thrive throughout North America. This afternoon, I found four kinds of clover along the Gull Harbor gravel road: White Clover; Red Clover, bigger blossomed and taller than White Clover, with leafy stems, blooming in reddish to purplish spheres of Sweet Pea flowers; Low Hop Clover, creeping along the ground with tiny, yellow, burrlike flower heads; and willowy White Sweet Clover, some of which grew over my head. If I hadn't read about it, I'd never have known the last one was clover at all.

❧ ❧ ❧

SWEET CLOVERS

White Sweet Clover doesn't look at all like low-growing White Clover, except for its three-part leaves. A common field and roadside plant, Sweet White Clover grows high and many-branched, sometimes over six feet, like an elaborate candelabra with thin, graceful (4- to 6-inch) columns of tiny, Sweet Pea flowers. Rich in nectar, White Sweet Clover is usually abuzz. Dried, it makes sweet-smelling hay. Yellow Sweet Clover, except for its color, looks just like it.

JULY 11

Common Mullein

I find it amazing that I can name so many plants and birds I see on my morning walks now. The ones whose names I don't know are the exceptions. I don't like to pick things anymore, except one sample so I can really study it, and then I check to be sure it isn't endangered. I knew that this morning's wasn't endangered—Common Mullein really is common— but I still felt terrible as I broke off the spike of an impressive five-foot plant, and even more terrible when I learned that Common Mullein takes two years to flower.

If I had left it there, it would have flowered most of the summer, the inch-wide, five-petaled, yellow flowers appearing only a few at a time along the many-budded spike. The plant lives only two years, its first year spent close to the ground as a roselike cluster of soft, furry leaves. You can often find these yearlings near one of their full-grown siblings.

❧ ❧ ❧

THE DISTINCTIVE COMMON MULLEIN

Look for a sturdy, waist- to shoulder-high plant, stalk surrounded with thick, velvety, lamb's ear leaves. The densely budded top blooms with several yellow flowers. Common Mullein is so tall that it often stands above other field flowers, making it easy to spot: Its range includes all of North America.

JULY 13

Yellow Goatsbeard

This morning at dawn, I came upon a weedy field completely blanketed with shining white gauze. Backlit by the rising sun, mist rising eerily above, possibly a million dew-studded spiderwebs glittered before me. Cobwebs had been worked over head-high Sweet White Clover, between every sumac, yarrow and milkweed, hung between knee-high spikes of Timothy, spun over roses and raspberries. Webs seemed slung in all directions, coming off the shoulders of sumacs like spokes of a wheel. The entire field was veiled above and worked below in knee-deep lace so dense that one step into it would have destroyed a half a dozen perfect specimens.

This extraordinary sight was edged in dazzling, delicate puffs, big as tennis balls, jeweled with dew. Some had been blown apart, the seeds shaped like gauzy, upturned umbrellas big as Daddy Longlegs, many caught like ghostly spiders in webs around them. Others plants were just beginning to open into spiky, yellow blooms. This, I discovered, was Yellow Goatsbeard, and I'd arrived at just the right time to see it at its best.

ᥬ ᥬ ᥬ

SPOTTING GOATSBEARD

It's easiest to tell goatsbeard after it's gone to seed: You can't miss the huge, dandelion-like puffs. Look for the yellow blossom in the morning, before it closes up at noon. It resembles a large, sparsely rayed dandelion, but spiky, green bracts extend beyond the rays. Long, grasslike leaves also help distinguish Yellow Goatsbeard from most dandelion look-alikes. Goatsbeard blooms all summer and fall throughout North America, except in the southeast corner.

JULY 14

Common Grackle

Iridescent, blackish birds that hang out in loud crowds have always been grackles or starlings to me, but I have trouble knowing which. Last spring I discovered a treeful of birds that I identified as starlings, but then I forgot the details. I never forgot their nonstop musical whistling, chortling and imitations of many birdcalls, though.

Early this morning, I was rudely awakened by an urgent scritching on the roof, accompanied by wild squawking and shrieking. I went outside to see what all the excitement was about and scared off, but not far off, about a hundred big, iridescent black birds with long, spoonlike tails. I decided these had to be grackles: Their noise did not at all resemble the circuslike silliness of the starlings I'd heard. A closer look at them through binoculars and a glance at the guidebook confirmed this.

The metallic green, blue or purple sheen on a grackle is not due to pigment, but to the reflection of sunlight off the scaled feathers. A grackle in the shade looks black.

JULY 16

Double-crested Cormorant

I've been wanting to write about cormorants ever since I saw them sweeping up the lakeshore last spring in big, black Vs, or strung out single file, but I've been waiting to see one up close. Here I see them fly high over the harbor in threes and fours. I'm told that last year about fifty cormorants hung out around the harbor, fishing around the docks, posing spread-eagled on the posts. This year, though, there's not a one. Some say they might come around later in the summer, when the nesting season is over.

I'm not going to wait. The cormorant's size and blackness give it away, even at a distance. *Cormorant* means "sea crow," and it's as easy to tell one from gulls, terns and ducks as it is to tell crows from sparrows and robins. Even in flight, the yard-long, pitch-black cormorant, its neck kinked back, head slightly raised, looks prehistoric, and up close, even more so. When pictured in guidebooks, the straight bill, hooked at the end, and long, black neck bring to mind a pterodactyl. Mary says the babies are even stranger; she's seen them featherless at several weeks, black skin baking in the sun as they sat in a big, messy nest made of sticks.

Cormorants are incredible divers and fisherbirds. In Japan, cormorants are leashed and used like hunting dogs to bring in fish, the neck rope preventing them from swallowing their catch. I'd seen that years ago on television, but I never knew we had cormorants here. We do. Huge clouds of cormorants cruise some parts of the Great Lakes, their capable fishing doing little to endear them to humans after similar prey. Now, not only can I tell one from a distance, but because I am inland, I'm fairly certain of what kind it is.

🐦 🐦 🐦

DOUBLE-CRESTED CORMORANT: THE ONLY INLAND SPECIES

A group of large, goose-sized, black waterbirds flying in a silent V (unlike Canada Geese, which honk loudly in flight) or single file are probably some kind of cormorant. Look also for a slightly elevated head and kinked neck. If you're inland, they are likely to be Double-crested Cormorants. If you're near salt water, they could be one of six kinds of cormorants, including the Double-crested. Up close, you can tell the Double-crested from the rest by the featherless, bright orange pouch under its bill.

JULY 17

Soapwort

The other day I was strolling along the harbor, noting the Sweet White Clover frilling high over the Blueweed, golden St. John's-wort, pink milk-weed and nodding white heads of Queen Anne's Lace, when I came upon a knee-high cloud of puffy, white flower clusters. I picked one—a sturdy stalk spiked with smooth, narrow leaves—brought it home to identify and ran into a problem: I couldn't find it in my guide.

This just didn't make sense. A flower this profuse had to be listed. Finally, I found it in Newcomb's guide (see Bibliography), which does not arrange its flowers by color, as most wildflower guides do, but by the number of petals on each blossom. My cluster of five-petaled flowers was Soapwort, also known as Bouncing Bet, which, the book noted, can be pink or white, and sometimes "double." Now I found Soapwort easily and in all my guides, but in the "pink" section.

Today, a patch of pink Soapwort burst into bloom on the other side of the road. On my way home, I passed Mary, who, seeing my pink sample of Soapwort, said that a double variety grew along Sloptown Road. I jumped in the car and found them there, which brought up another problem: Ordinary Soapwort has five petals, spreading out from a long, green tube, something like Bladder Campion. Each "double" Soapwort flower, how-ever, was fluffed with at least twenty petals. "Double," it seems, actually means "ruffled" and cannot be taken literally. I think it amazing that in just three days I found three variations of one plant.

Soapwort, although poisonous when ingested, whips into suds when the stem and leaves are bruised and soaked in water. It's reputed not only to clean fabric, but bleach it. I myself would rather buy my soap than stalk this wildly beautiful detergent.

ð. ð. ð.

CLOUDS OF WHITE OR PINK

A cloud of white or light pink blossoms atop one-to-three-foot spear-leafed stalks is, at least this time of year, probably Soapwort, also called Bouncing Bet. Soapwort spreads quickly on a vigorous, underground root system, so the plants are usually found in large groups, On "single" varieties, check for five notched petals spreading from a narrow, green tube, and take time to smell the strong, sweet fragrance. Soapwort is found throughout the United States and in parts of southern Canada.

JULY 19

Green-winged Teal

On my way home from Sloptown Road yesterday, I stopped at Barney's Lake, the place I'd gone to find dragonflies. The small lake lay flat and dark beneath a wide ribbon of reflected oaks and cedars. The marshy edges bristled with bulrushes, up to the little sandy beach where I stood. Multi-hued wildflowers were thick back to the road. I thought I might find warblers there—I'd seen a Common Yellowthroat last time—but what I found was a middle-aged man eating popcorn out of a white paper bag. "You a birder?" he asked, glancing at my binoculars. I confessed. "There're two loons out there," he said. I checked two dark dots across the lake with the binoculars, and he was right. Then he pointed out a Great Blue Heron standing on one leg in the middle of some ducks that my binoculars were not strong enough to let me identify.

The man went to his truck and screwed a scope onto the half-rolled-

---

PUDDLE OR DIVING DUCK?

Puddleducks, sometimes called "dabblers," usually tip, tail up, to feed just under the surface, eating mostly plants. Pintails are puddleducks, as are, among others, mallards, Wood Ducks and all the teals. *Diving ducks* dive and swim under the surface to catch food, which tends to be more animal than vegetable. Scaups, mergansers, Buffleheads and Goldeneyes are some examples of diving ducks.

---

down window of his truck. I noticed an issue of *Birdwatcher's Digest* lying on the front seat. He glanced through the lens. "Green-winged Teals," he announced. Taking my turn, I saw four small brown ducks with emerald-green wing patches, huddled beneath the tall, fierce Heron, yellow and white lilies floating in front of them, slender, dark bulrushes bowing behind. They were females. The light gray male, with his bright, rust-colored head and iridescent slash of green across his cheek, was absent.

I wish I'd seen the ducks take off. Although one of the smallest puddle ducks (twelve to sixteen inches), Green-winged Teals are famous for their tightly packed squadrons that fly at wicked speeds, like Blue Angels, in perfect harmony, changing direction simultaneously, as if by remote control.

                         ❧ ❧ ❧

MORE THAN ONE GREEN-WINGED DUCK

Several ducks have a green wing patch, including the Cinnamon Teal and even the Blue-winged Teal, but their green patches are preceded by another patch of light blue. The Green-winged Teal's green wing patch is the only wing patch it's got. Look also for the male's vertical, white stripe just back of his black-speckled breast, and the horizontal, green, paisley-shaped stripe on each side of his head.

JULY 20

Spotted Knapweed

I'm enjoying the recent explosion of little, hot-pink flower puffs among the frilly Sweet White Clover and Queen Anne's Lace. They are Spotted Knapweed, a tough, wiry plant that grows several feet high, branching all over the place, stems tipped with cherry-sized pink explosions. Right now I'm seeing Spotted Knapweed everywhere. I tried to pick one and almost cut my fingers on the rough, tough stem. I had to return with a scissors.

Knapweed is another one of those flower heads, each "petal" a separate flower. The rays burst, thistlelike, from a hard, knobby green base; supposedly, *knap* comes from the Old English word for "knob."

ᴥ ᴥ ᴥ

LOOK FOR A SPINELESS "THISTLE"

Knapweed—there are many kinds—blooms from a hard, round base and looks like a sparsely rayed thistle, but its thin, deeply lobed, gray-green leaves are very unlike a thistle's spikey foliage. In addition to the leaves, you can distinguish Spotted Knapweed from the other knapweeds by the little black dots on the "knob." Spotted Knapweed is found all over the United States and in parts of southern Canada.

JULY 22

Redtop

Many of the grasses are now in full flower: tall plumes arching gracefully over lush green foliage; slender, cattail-like tops bumping each other in the breeze; flat, narrow, basket-woven tops; delicate, frilly tops; short, lacy flower heads on threadlike stems. The grasses form an exquisite, swaying screen through which to view the sailboats, but I still can't name any of them but Timothy.

I've tried. I called the Central Michigan University lab station here, asking if I could please bring over about ten grasses to be identified. I was sent to their "herbarium" to look at pressed samples and match them up with my own. An herbarium turns out to be a library of specimens kept, not on display, but in arranged cabinets by name. Trying to identify a grass by looks was like trying to find a book in a library by the picture on the cover, so I was back where I started.

Yesterday, I gashed my hand on a grass blade when I tried to pick it and dripped blood all the way home. I used a scissors to cut a sample of each of the rest of the most common and beautiful species and decided that I would learn what they were if it took all day. It did. I stared at eight grasses for three hours. Each was fairly distinctive, but I still had difficulty seeing the differences. I paged through my books for another three hours. I identified, as best I could without a microscope, five. Grasses are classed by chromosomes and characteristics not visible to the naked eye, so a guess is the best a casual observer can do. The surest of all my guesses was Redtop.

Clumps of Redtop grow here along the shore, among the delicate, blue Brook Lobelia, Timothy and clover. The lacy, purple-red grass flowers bloom on threadlike stems of thin stalks about calf high. Although not tall, the pyramid-shaped flower head is big as a fist. Redtop, I discovered, is one of the most important perennial grasses in the United States, making great hay, forage for animals and, mixed with other grass seed, lawns. So not only is there lots of it around, it's fairly easy to spot.

❧ ❧ ❧

A RED-HEADED GRASS

The relatively bright color of the flower heads helps you see Redtop among the greener grasses. Look for a twelve-to-eighteen-inch grass with thick, long, harsh-feeling, basal leaves and a red-to-purple, delicate top with tiny, seed-sized flowers on thread-thin stems. The beautiful whorls formed by the lower branches on the flower head can help you tell Redtop from look-alikes. Redtop grows in wet places and dry, throughout the United States.

JULY 23

Barn Swallow

The afterglow of last night's sunset turned the whole cloud-scudded sky flamingo, so I walked out to the dock at the abandoned Coast Guard station, the better to watch the nearly full moon slip in and out of the pink, wisping clouds. As I passed close to the empty, white frame building, I was narrowly missed by swallows darting in and out of a broken window. Loud cheeping echoed inside. I guessed that they were Barn Swallows, the swallows that prefer the cushy, indoor life, building their cup-shaped, mud-and-straw nests in (or sometimes on) structures like barns, sheds, abandoned buildings and boathouses. Their unusual tails clinched my identification.

As the evening colors intensified, the only birds I saw were diving terns, ducks flapping past low and fast and Barn Swallows swooping and skimming the riffled harbor. It was late in the evening for such small birds to be up, but Barn Swallows sometimes cover six hundred miles a day catching insects for their young, flying from sunrise until the bats, which in the dark they resemble, come out.

ᏧᎬ ᏧᎬ ᏧᎬ

THE ONLY SWALLOW-TAILED SWALLOW

A darting, swooping, 6- to 8-inch bird with a deeply forked tail is likely a Barn Swallow, the only swallow with a true swallowtail. It's also the only swallow with two white marks on the tail and a bright orange breast. The hooded head and upperparts are dark. The Cliff Swallow looks similar, but its tail is notched, not forked, its breast is white, and it lives outside of structures, not in them.

JULY 24

Quackgrass

I have been on Beaver Island nearly two months now and my life has settled into a delicious routine. I sleep until I am wakened by noisy birds, or by Jon's truck roaring up the long driveway as he arrives for his morning coffee at Don's tiny one-room cabin next door, or by the ferry boat's resonant blasts that daily announce its eight-thirty departure. Then I get up. Before long I am wandering sleepily down to the water with a mug of hot coffee. There's a rock on the shore there, located just so in a patch of sunlight, tree shadows falling to either side, where the rising sun warms my back and lights up the town across the harbor with a scoured radiance: the church steeple, the flat, painted storefronts, the yachts and sailboats anchored in the marinas, the small amphibious plane floating near a harborside motel.

After a while, I stop by Mary's to tour her brilliant, bursting garden for the latest rose debuts, after which I bicycle the harbor road a mile to town, pick up my mail at the white-frame post office and stop by the St. James Boat Shop for more coffee. I am welcomed there by Bear, a big black lab who leaves sawdust pawprints on the shoulders of my T-shirt when I settle on the tall stool at the large workbench in the middle of the room. Bill, the lanky, white-haired man on the other side, is probably gluing together the slats of a wooden bucket or working on a three-foot model canoe made from thin slices of cherry or birch. Huge planks are stacked along three walls; buckets in various stages of construction await the next step; full-size wooden canoes, resting on sawhorses, glow with craftsmanship, drawing the immediate attention and awe of the tourists who wander in and out.

By about ten o'clock I start working. Working often means spending an hour or two wandering through the woods, along the shore, or driving to more remote parts of the island, after which I come home, select my topic for the day and write about it. By four o'clock, if I haven't had too many visitors, I go swimming off a small wooden dock in water so clear I can see my pink toenails through water up to my neck. Sometimes I'm not finished so early. New friends wander in and out, show up to chat, drink coffee. The over three hundred Beaver Island inhabitants are friendly folk, and the population explodes to over two thousand in the summer. I have decided to enjoy the interruptions rather than get sticky about disturbances to my work. In Douglas I was rarely disturbed by anyone, and I find myself delighted by all this unanticipated good company.

I have a telephone, but no television. I don't miss it in the long days of summer, when sundown and bedtime often occur simultaneously. Here I spend the evening hours dropping in on friends, dining or dancing at the local bar and grill, reading, or sketching. I almost always go for sunset walks.

It was on yesterday's walk, swallow-watching by the lighthouse, that I found a third fairly easy-to-name grass. Fortunately, this one differs greatly from the taillike Timothy and the delicate Redtop: Quackgrass is a relative of wheat but lacks the "beard," or long hairs. About as widespread as Redtop, Quackgrass seems to hit the other end of the value scale: My books described it as "noxious," "hard to get rid of," and "the worst grass weed in the United States." Since I'm not fighting the stuff, though, I find it beautiful.

ન& ન& ન&

HERRINGBONE GRASS

To find Quackgrass, look for a fairly flat flower head with oval seed shapes going up each side of the half-to-three-foot stem in a herringbone pattern. The pattern is precise and regular, looking woven, and lacks any hairs or "beard." Quackgrass is common in the northern half of the United States and in Canada.

Yellow Woolly Bear

When I first moved into my sixteen-by-twenty-foot cabin, I thought I might suffer acute claustrophobia; instead, I feel quite cozy here, having found room for everything. Well, not quite everything. Last year the island garbage dump was closed by the state, replaced by a building called the Transfer Station. "Garbage" has become a collective noun in a new sense: I must share my intimate quarters with at least eight containers: Plastic milk bottles, cans (tops and bottoms removed, washed and smashed), glass bottles (divided into clear, green and brown) and nonrecyclable refuse (put in green plastic bags that cost a dollar each) must be dropped at the Transfer Station for recycling or shipping to the mainland; aluminum cans and refundable bottles go to the grocery store; paper is burned on a windless day in a rusted barrel; raw vegetable and fruit leavings are collected for my friend Mary's chickens; and leftovers go to the sea gulls.

Talk around town has it that since the dump closed, the sea gulls, which would rather steal a meal than catch one, have been mighty hungry. Some are reported to have snatched chicken pieces right off an outdoor grill, or even from a poised fork. So, at Mary's advice, I feed the sea gulls. All this is to explain why I was dumping the last of a tiresome casserole on the beach, which is what I was doing when I spotted a fuzzy, rust-colored caterpillar munching on some nearby White Sweet Clover. I picked the branch, brought it home, and put it in a jar with the caterpillar. This morning the caterpillar was still there, clumps of little hairs bristling all over it, but the clover—stem, leaves and flowers—was as completely devoured as yesterday's leftovers. Caterpillars are big eaters.

I thought it was a Woolly Bear, the caterpillar for the Isabella Tiger Moth, but this caterpillar lacked the Woolly Bear's black front and back. There are many kinds of tiger moths—named for their black and yellow stripes (although some, including the Isabella, are fairly nondescript)—all of which have hairy, bristly larvae, i.e. caterpillars.

It's hard to identify caterpillars: They are often very similar, and moth and butterfly guides don't often show each species' caterpillar stage, so I'm

---

CATERPILLARS HAVE ONLY SIX LEGS

Like the moth or butterfly, caterpillars are insects and, despite appearances, have only six legs. A caterpillar's real legs are clawlike structures up by the head. The others are false legs, called *prolegs*, used for motion and glomming onto vegetable prey. Whatever their varying sizes, all caterpillars have thirteen body segments (the last two are joined) and ten pairs of "nostrils," or breathing holes, along their bodies.

---

guessing that mine was a Yellow Woolly Bear, which will become a Yellow Woolly Bear Tiger Moth. I was surprised to learn that whatever it was, it shared with others a number of caterpillar characteristics.

Most caterpillars are smooth, and most spend the winter in cocoons. Not so the bearlike tiger moth caterpillar. My Yellow Woolly Bear, found in midsummer, is likely a first-generation caterpillar: it will spin a cocoon, and emerge a tiger moth, which will lay the eggs for a second generation of caterpillars. These will hibernate in their fur coats, under rocks or logs, waiting until spring to spin their cocoons.

ха ха ха

WHEN BEARS HAVE WINGS

A fuzzy caterpillar, especially one that curls up when disturbed, is probably the larva of some kind of tiger moth.

JULY 27

Ruffed Grouse

My neighbor returned from a trip to the mainland bearing a sack of ripe peaches, promising a peach-raspberry cobbler if I'd come up with the berries. Minutes later I was padding through the woods. It took me half an hour to get a good cupful—the season had already peaked—and I found myself standing several times in lush Poison Ivy, but it was wonderful out there under the sun-warmed cedars, reaching through rose and raspberry thorns and stepping through the prickly juniper.

As I ambled home through the woods, wondering what I could possibly write about raspberries, sharp, catlike screaming and a sudden bluster of wings stopped me—and nearly my heart! A chicken-sized bird roared up from almost underfoot and began dragging its wing on the ground in front of me, tail fanned out like a turkey's.

Ha! I thought smugly. The old broken wing trick.* You can't fool me! I began poking around for the chicks that were surely hiding somewhere nearby, but I couldn't find them, nor could I possibly hear cheeping over Mama's shrieking. At last she ran off into the woods, although not far, and wait as I would, I could hear no chicks. She had fooled me after all. (The cobbler, however, was delicious.)

Being outwitted by a Ruffed Grouse, a ground bird famous for some fairly dim-witted behavior, is pretty embarrassing. I haven't seen one fly into fences or buildings, as some reports claim, but last summer a Ruffed Grouse stopped my van cold in the middle of the road. Having honked to no avail, I finally had to get out of my car and run it off.

Not realizing that there are many kinds of grouse, I guessed the one that startled me was a Ruffed Grouse; however, the Sharp-tailed Grouse and the Spruce Grouse share a good part of the Ruffed Grouse's mostly northern U.S. and Canadian range. In the Rockies or the Northwest, it might have been a Sage, Sharp-tailed or Blue Grouse. In certain parts of the Midwest, it might have been a Prairie Chicken. But although all grouse are ground birds with little heads on plump, brownish bodies, their tails are the giveaway.

ஐ ஐ ஐ

GROUSE TAILS

A chicken-sized gray-brown or reddish ground bird with a many-banded fan tail is probably a Ruffed Grouse. Here are some others:

---

* Several species of birds, including Killdeer and grouse, may distract attention from their young by pretending to suffer a broken wing.

GROUSE	TAIL
Blue	Black with gray band at end
Spruce	Black with rusty band at end
Sharp-tailed	Pointed with white sides
Sage	Long and pointed, a spiky fan on display
Prairie Chicken	Small black fan tail

A RUFFED GROUSE IN A PEAR TREE

A friend here told me he saw a partridge in his pear tree last Christmas Day. What he was referring to was a Ruffed Grouse, which in North America is also known as a partridge. "Partridge" does sound so much better than "a Ruffed Grouse in a pear tree." Perhaps the English partridge is more musical than our male Ruffed Grouse, which is famous for his thunderous spring ardor, a drumming made by a trick of the wings that is said to sound like a chainsaw or large farm machinery.

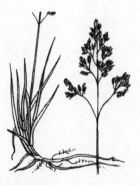

JULY 29

Sweetgrass

Today a friend gave me a thumb-thick braid of fragrant grass, knotted at both ends. It was Sweetgrass. Looped over a nail above my computer, it scented my cabin in minutes. All summer I'd been hearing about Sweetgrass—always spoken of with reverence—but no one had shown me any, or told me how or where to find it. I was never sure if they couldn't or if they held the location too dear to reveal. "It looks like other grass," I was told, "but you can tell it by the smell." So far, all the grasses I'd found smelled like grass. My friend Eric, however, said he'd show me a patch of it.

Eric, who seemed more at home outside than in, said he associated Sweetgrass with clean water, unpolluted places, fresh air and woods. To find it, we proceeded very slowly through pillared woods on deeply layered, rust-red cedar lace. Eric taught me how to see with my feet, a slow, quiet

---

PUT EYES IN YOUR FEET

If you put your weight down on the outside edge of your foot (instead of your heel) and roll your foot slowly toward the arch, you can feel what is underneath and withdraw, if need be, freeing your eyes to look up and around you and quieting your progress. Walking normally—landing on your heel and rolling forward—your step is committed, you land hard, and instead of looking up and around you, you must look down for obstacles.

---

walk that woke up my senses, took the bumble out of my step and made me feel more a natural part of my surroundings.

We found the Sweetgrass growing between coarser grasses, wildflowers and small Smooth Rose shrubs in a boggy, damp depression in the soft ground. From each single, delicate stalk flowed a long, limp tail of three or four thin blades, a foot to three feet long. The blades, arching gracefully toward the ground, smelled strongly of vanilla.

Sweetgrass is also called Holy Grass, valued for Native American ceremonies and as a symbol for many good things, among them purity and sweetness of spirit.

&a. &a. &a.

RECOGNIZING SWEETGRASS

It isn't easy to find Sweetgrass, growing as it does among other grasses. Look for three or four long, glossy, arching blades that are dull underneath. In spring, a handsome, shining brown *panicle* (cluster) of delicate small "flowers" on threadlike extensions helps identify it. Although not particularly common, Sweetgrass ranges throughout the northern third of the United States and in much of the West, including the Rocky Mountain states.

JULY 30

Bigtooth Aspen

I've been hearing Islanders refer to "popple" trees here, and it took me a few weeks to figure out that they were referring to the two kinds of aspens that grow here; aspens, it turns out, are a kind of poplar. I've recognized Quaking Aspen ever since I lived in the Rockies, where its one-to-three-inch ace-of-spades leaves shimmer in the slightest breeze—silvery in summer, molten gold in fall—among the dark evergreens. In the Northeast, Bigtooth Aspen grows as well, with a leaf twice as large and egg-shaped, with a seashell, scalloped edge. I think Bigtooth Aspen leaves are among the prettiest I've ever seen. Bigtooth Aspen thrive behind my cabin, their fancy foliage strewn on the forest paths. The bark on younger aspens (both Quaking and Bigtooth), almost as smooth and light as Paper Birch, stands out in the cedar groves where they are often found, but aspen bark does not shred or peel as Paper or Yellow Birch bark does.

❧ ❧ ❧

DECORATIVE BIGTOOTH ASPEN LEAVES

Find a medium-sized (3- to 6-inch), egg-shaped leaf with nine to fifteen big, dull teeth up each side, and you have a Bigtooth Aspen leaf. The teeth on Chestnut and beech leaves are wide, too, but sharp, and the leaves are spear-, not egg-shaped. The edges also help distinguish aspen from birch: Birch leaves are doubly serrated, while aspen leaf edges have single teeth.

WHY ASPENS QUAKE

I've always wondered why both kinds of aspen leaves quake more easily than other leaves. The answer lies in the stems: Not only are aspen leaf stems very long—on Quaking Aspen, even longer than the leaf—but they are thin and flattened vertically. The long stem gives the leaf freedom to flutter, its flat surface a sail for even a breath of moving air.

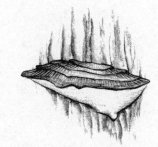

August

Indian Pipe

Today is Mary's birthday, so this afternoon I decided to make her a chocolate cake. Lacking an electric mixer, I sat on my front stoop, shaded by oaks, yellow bowl in my lap, beating the batter with a wooden spoon, counting the strokes out loud. I borrowed a pan from next door, poured the batter in and put it in the oven, which I had not used until now.

While the cake was baking, I thought I might duck back in the woods for a few minutes and pick wild raspberries to sauce it with. I'd been in the tree shadows only minutes when I came across a plant so strange that even though I'd never seen one before, not even a photograph, I knew what it was. I must have read the name somewhere, though, because when the ghostly stalks appeared, like white asparagus with nodding, bell-shaped heads, I immediately thought "Indian Pipe." The name perfectly described each stalk, which resembled a long-stemmed pipe stuck in the ground. About ten of the eight-inch stalks emerged separately from the ground, crowding together, bent in the same direction.

Later I read that some Native Americans once believed that seeing the ghostly-looking, white, translucent plant was bad luck, which may account for my having gotten confused in the woods, taking a wrong turn and having to rush breathless back, tearing down this path and that, making it back just as the cake began to burn. Indian Pipe does not need magical powers, however, to qualify as an odd plant. Lacking chlorophyll, it resembles a mushroom more than a wildflower and gets its nutrients from fungi in the soil. It even feels cool and clammy. I didn't see any other flowers in the woods; most midsummer ones seem to seek the sun.

≈ ≈ ≈

A FLOWER THAT LIKES THE SHADOWS

Look for Indian Pipe in deep woods, especially (but not only) in conifer woods, all summer, anywhere in North America. Unequipped to turn light into food, Indian Pipe thrives in the shadows. From two to twelve inches tall, the waxy, translucently white stalks grow in small groups, each one turning black when the nodding, bell-shaped flowers are fertilized. *Do not pick!* Indian Pipe is a protected plant in many states.

AUGUST 2

White Spruce

Nature-savvy friends, who are often happy to share their superior knowledge, can save an amateur like me much time and confusion. Roy, an ex-forester friend, today taught me some handy basics about the common trees of the island: The most common island oak is Red Oak; the pine is White Pine; the fir is Balsam Fir; the hemlock is Eastern Hemlock and the spruce is White Spruce. If I can tell a pine from a fir from a spruce from a hemlock, which by now I can do pretty well (see page 43 if you can't), I can identify most of the island's evergreens. The one I was most grateful to know was White Spruce, which I've never been able to tell from Black Spruce.

White Spruce grows so lush and blue-green here that I almost mistook it for Colorado Blue Spruce. White Spruce cones are half to one inch long, and the needles are about the same length, spiraling the twig but crowding on the upper side. In most of the White Spruce's range—mainly the northern part of the Northeast and all of Canada—the only other similar spruce is the Black Spruce. As you near the Atlantic coastline, however, you get Red Spruce, too. Telling these three apart is not simple: The needles and little woody cones dangling from the branches are very similar—between a quarter and one inch in length.

❧ ❧ ❧

THE RED, THE WHITE AND THE BLACK SPRUCES

If you think you're looking at a Red, White or Black Spruce, but you don't know which, try the twig test: If the twigs are hairy and you're not in Red Spruce range (the far

Northeast), it's Black Spruce (Red Spruce twigs are a little hairy); if the twigs are smooth, it's White Spruce. Or try the sniff test: White Spruce (also known as "Skunk Spruce") supposedly smells like skunk (mine smelled like Christmas), Black Spruce like menthol and Red Spruce like an apple or other fruit.

AUGUST 4

Striped Maple

Yesterday's island tree tour included a hill that was dense with the slim trunks and enormous leaves of Striped Maples. I don't remember seeing a Striped Maple tree before, but then, I probably wouldn't have recognized it for a maple, since the leaves don't conform to the leaf shape on the Canadian flag. If I'd known then what I know now, I at least would have recognized a maple when I saw one, whatever kind it was.

I've been uneasy about getting into maples; to my untrained eye, maples, like sparrows, all look alike. Meeting the Striped Maple has given me the courage to begin it, being one of the easiest maples to identify.

A TRICK FOR RECOGNIZING A MAPLE

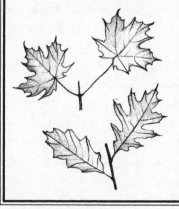

Maples have a distinctive feature for trees with large, broad leaves: Their leaves are *opposite*, growing in pairs opposite one another. The leaves of most species of big-leaved trees, such as oaks, basswoods, hawthorns, elms, beeches, poplars and birches, are *alternate*, growing in staggered fashion.* You can tell a maple leaf from the similar sycamore leaf in this way, too, sycamore leaves being alternate. (Sycamore bark is also very light.)

* The catalpa is an exception, also having opposite leaves.

&a &a &a

A TREE OF A DIFFERENT STRIPE

A maple in the northeast quarter of the United States with light stripes on a slim, olive-colored trunk is likely a Striped Maple. No other tree has such a trunk. Then check for hand-sized, crepey, three-lobed leaves with doubly saw-toothed edges. Although usually a small tree, the Striped Maple boasts the largest leaves of any maple species. Mountain Maple and Red Maple, two maples sharing the same range, have similarly shaped leaves, but the bark is not striped.

AUGUST 5

Red-backed Salamander

I've been making baskets from cedar bark, a skill learned from Eric, and the resulting containers, squat and organic-looking, so enchanted me that today I took to the woods to find a felled cedar log I could debark for another. During my search, I saw a female Baltimore Oriole and exchanged meows for a full minute with a curious Catbird before coming upon a large cedar log on top of a pile of "popple." While stripping its bark, I uncovered a beautiful surprise: Two tiny salamanders, maybe two inches long and brown, with a wide red stripe from head to tail, scurried between the bark and the naked log.

If I'd known that salamanders are not dangerous to people, I'd have scooped them up and taken them home among the raspberries I'd been picking along the way. I've since discovered that I was lucky to find them: Salamanders, although common, are nocturnal, hiding in and under logs and beneath rocks during the day. Multicolored in bold red, green, yellow and orange, polka-dotted and striped, salamanders seem to me to be the tropical fish of the forest. They even like moist places.

My salamanders were easily identified as Red-backed Salamanders,

---

FINDING THE RED-BACKED SALAMANDER

To find salamanders, look under rocks and logs in damp places. A two-inch brown or gray salamander with a wide red stripe down the whole top of its body is probably a Red-backed Salamander. Also common is its morph, called the Lead-backed Salamander, which is uniformly gray.

---

one of the most common salamanders in the northeast quarter of the country. The Red-backed Salamander grows no longer than two inches and, unlike most salamanders, possesses neither lungs nor gills, breathing instead through its skin and the lining of its mouth.

🐸 🐸 🐸

SALAMANDER OR LIZARD?

Look at the skin: A lizard-shaped creature with four splayed legs, a long tail and damp, smooth skin is a salamander. Salamanders are amphibians—damp skinned, four-legged vertebrates (animals with a segmented backbone)—while the similarly shaped lizards are four-legged reptiles—scaly vertebrates. Having little outside protection, salamanders need damp places to keep from drying out.

AUGUST 6

Artist's Fungus

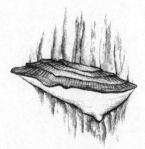

I got so excited about salamanders that today I went hunting for more, overturning rocks and logs back in the woods. I uncovered earthworms, crickets, ants among ricelike eggs, rusty creatures scurrying on many legs, but no salamanders. Poking among so many cut and fallen trees, I began to notice the many fan-shaped fungi growing on the dead wood. I pulled off one as big as the stretch of my hand. It resembled a hard, dry horse's hoof, furrowed gray-brown on top, white and paper-smooth un-

---

DRAWING ON MUSHROOMS

If you find a fresh Artist's Fungus—the white surface underneath will be slightly damp—you can draw pencillike lines on it with just your fingernail or a twig. If you've broken the fungus off, it will harden, holding the picture. If you've left it on the tree, your picture will enlarge with the mushroom, which can go on growing for years, sometimes reaching twenty inches!

---

derneath. When I got it home, however, the slightly damp, white under-surface was bruised brown, holding my prints as if my fingers had been inked. When I identified it as an Artist's Fungus, I knew why.

Most shelf fungi (not all) are too woody to be very tasty, so are often not well represented in mushroom guides. I find shelf fungi very interesting, however, and easy to find, often popping out right at eye level on tree trunks. Roy claims that the roots of some shelf fungi can grow to six feet inside a tree, sometimes threatening a live tree. More often, they serve to decompose the dead ones.

ɛ̀ə ɛ̀ə ɛ̀ə

FINDING ARTIST'S FUNGUS

Look for a furrowed, brown or gray-brown shelf fungus shaped like a fan, sometimes like a horse's hoof, with a white undersurface (pores). Sometimes the white becomes bruised with brown smudges or bruises. The Redbelt, a similar but darker shelf fungus in much the same range—throughout the United States (except the deepest South) and southern Canada—can be distinguished by the red furrow around the outside edge.

AUGUST 9
Daisy Fleabane

For a couple of weeks now, bushy plants bristling with tiny daisy-like blooms have been crowding the Gull Harbor roadsides, practically growing in the gravel. I identified them as Daisy Fleabane, but didn't get around to writing them up until this morning. Needing a sample, I popped out to the road outside my cabin and picked one. When I started reading up on fleabane, though, nothing was making much sense. My Daisy Fleabane just wasn't the Daisy Fleabane in the book. I looked it up in another book, thinking there were probably several kinds of fleabane, and there were, but it still didn't check out: The "petals" didn't count out right, the leaves were too small and the stems weren't hairy.

I didn't know if I'd made a mistake the first time I identified the plant, or if today I had a different but similar plant, so I went back out Gull Harbor Road to look again. There it was: Daisy Fleabane, hairy stems and all. What I had found at the end of my driveway was a Bush Aster, and this is just the time of year to start mixing the two plants up; fleabanes bloom all summer, but asters are more a late summer and fall flower. Now that I know what to look for, I won't make that mistake again.

Despite the name *bane*, which means poison, fleabane wouldn't hurt a flea, although folk once thought it would. It's a common plant all over North America and it blooms from June to October, so finding some should be easy.

❦ ❦ ❦

HOW TO TELL FLEABANES FROM ASTERS

Look at the stem: Fleabane stems are hairy, especially lower down; asters have smooth stems. The petals are another good clue: Fleabane petals are short—often shorter than the disk diameter—thin and too numerous to count (40 to 100); at night, they close up over the yellow middle. Aster petals are wider, longer and fewer. (See also page 244.)

FINDING FLEABANE

Look for a small (¼- to 1-inch) daisylike bloom with a short fringe of very fine, white, pinkish or bluish rays around a yellow disk. There are several kinds of fleabane, among them Daisy Fleabane, bushing out atop a long, sturdy stem crowded with half-inch blooms, several per stem, and Common Fleabane, with larger and fewer blooms.

AUGUST 11

Purple Loosestrife

For the first time this summer, I left Beaver Island for a few days, one of which I spent canoeing with a friend through some Kalamazoo River wetlands near Saugatuck. As we put in, we were immediately overwhelmed by the huge bands of vivid purple-pink along the riverbanks, laid against the summer green as if by the long, loose stroke of a loaded brush. The color continued in both directions, as far as we could see, along the water and in it, crowding small islands. Who but an artist would be so bold? Surely we had somehow paddled into a painting. As we glided upriver, the color stretched on, too immense to feel real.

It was real, unfortunately. We were admiring Purple Loosestrife, a runaway, thriving import that not only crowds out native plants, producing as many as two million seeds per plant every year, but does little to feed wildlife, loved only by bees, bugs and butterflies for its abundant nectar. Like the Mute Swan, Purple Loosestrife is so aggressive that its beauty faded a little in my heart once I had become aware of its price. Brochures have been printed on its control, articles written. Experts concerned about the wetlands of the Canadian and U.S. Northeast are fighting to contain it.

On the river, I didn't know this. I thought the plant, blooming thickly on foot-long spikes atop a sturdy, five-foot bush, was Fireweed until I looked it up, but this was a case when the habitat helped identify the plant for me before I even got close enough to count petals.

        ঽ ঽ ঽ

LOOSESTRIFE OR FIREWEED?

Note where it grows. Purple Loosestrife likes wet, boggy places, blooming all summer in huge patches of solid color. Each bright pink-purple, ½- to ¾-inch flower can have three to seven narrow petals, usually six. The plant is found in the eastern half of the United States and Canada. Fireweed, found just about everywhere in North America, also blooms bright pink on tall wands and also forms large patches, but it prefers dry places: roadsides, clearings, and burned-out sites. Fireweed blossoms have four rounded petals.

AUGUST 12

Cardinal Flower

Like a lover who delights in delighting, the river yesterday seemed to take pleasure in presenting me with one fantastic new wildflower after another: blue, wine-colored, baby pink, white, most of them small flowers on spikes, but culminating in a tall explosion of blood-red blossoms. "Cardinal Flower!" I shouted, even though I'd never seen one before. The Cardinal Flowers appeared in groups along the riverbanks, in shaded, tree-thick places, the only red flowers we saw.

The Cardinal Flower is like the Scarlet Tanager: If you've ever heard the name, you'll know it when you see it, and you'll never forget the time and place of your first encounter. A kind of lobelia, each tubular Cardinal blossom flowers a little like a sharp-petaled violet, two petals on top and three on bottom. A slender red tube containing the stamens extends between the top petals.

The Cardinal Flower blooms June through September throughout all

North America except the western third, but it's been picked so much that it's no longer common. It takes wonderfully to cultivation, though, and seeds are often available.

ও ও ও

SOME SPIKY RED WILDFLOWERS

The Cardinal Flower, with its spear-shaped, toothed leaves, is not the only example of bright red blossoms on a spike. Standing Cypress, an eastern U.S. plant, also blooms red on long tubes, but in star-shaped bells. Tropical Sage, found in the southern quarter of the country, looks a little like the Cardinal Flower but has heart-shaped leaves. The western Towering Louse-wort has beak-shaped flowers, without separate petals.

AUGUST 13

Pintail Duck

My binoculars dug rings around my eyes, I watched so many birds on that Kalamazoo River trip. It pleased me that I could now recognize so many birds—chattering kingfishers, king birds stationed on dead branches, a Red-tailed Hawk, several Great Blue Herons, some large waterbirds that might have been Green or Night Herons—but by far the most numerous were ducks. We'd come around a bend, paddling quietly, and about fifty yards up a flurry of ducks would leap into the air, flap over the treeline and re-situate upriver. A few minutes later, we'd round another bend and the excitement would begin again.

We were never able to get very close, and despite their large numbers, I found it difficult to spot the ducks in the water. They seemed to melt into the shadows. Now and then a duck would let us get quite close, but I wouldn't see it until it flew up in alarm. By the fourth turn in the river, I found a straggler that wasn't yet on the wing. In the minute before it fled, I noticed its gray back, its brown head and the white on its breast that extended gracefully upward, like a finger, behind the bird's "cheek." That distinctive marking, plus the birds' behavior, identified this flock as Pintail Ducks.

I'd never seen a Pintail Duck before, but it turns out to be one of the most abundant ducks in North America, especially in the West. It is, however, one of the wariest ducks, taking immediate and graceful flight at the slightest disturbance. Except during nesting season, Pintails hang out in large groups.

It never occurred to me, out on the river, that I was looking at the wrong end of the bird. I missed the Pintail's most distinctive feature. Although Pintails can dive and swim underwater, they are described as puddleducks.

🐦 🐦 🐦

PINTAIL: CLUES AT BOTH ENDS

If you can't get close enough to a large, mallard-sized (20- to 30-inch) duck to see the head details, look at the other end: A male Pintail's black, narrow tail is as startlingly long and unlikely as the nose of a Swordfish. The male's head is also distinctive: The white upper breast extends up behind the face in a white crescent. The female is mottled brown, resembling many other female ducks.

AUGUST 15

Wood Thrush

On the last morning of my visit to Saugatuck, I heard a loud, fluting three- to five-note bird-song rising and falling, silvery, liquid and enchanting. "Listen!" I said to Ellen, with whom I was staying.

She put down her morning coffee. "I've never heard that before," she said. "Tell me what it is!"

I was sure it was a Wood Thrush. It could have been a Hermit Thrush, which also has a beautiful, fluting call, but the Wood Thrush is really the only thrush that can deal with human habitation. It is fairly common all over the eastern half of the United States and southeastern Canada. It's probably been singing in Ellen's woods for years, but since she's not in the habit of sorting out bird voices, she just heard the overall symphony.

Every time I hear the Wood Thrush sing, a thrill runs through me. I first heard it last spring in the early morning woods, then outside my bedroom window on the lakeside, and this summer outside my cabin. I knew what it was because I listened to a bird-song tape, I was so eager to identify it. I saw a Wood Thrush once, too, when one met its demise against a window pane of mine last May, but I have never seen a Wood Thrush sing.

I was waiting to write about the Wood Thrush until I saw one sing, but the season is getting so late that this may have been my last concert for the year. It's easier to identify by ear than by eye anyway.

☙ ☙ ☙

LISTEN FOR THE WOOD THRUSH

The liquid, flutelike notes of the Wood Thrush's call identify it easily. Heard most often in early morning and in the evening, the song can go on and on, rising and falling in three- to five-note phrases that sound a little like *Ay-oh-lee! Ay-oh-lee-ah (trill)*, etc., sometimes ending in high trills or buzzes.

FIVE LOOK-ALIKE THRUSHES

A brown-to-rusty–backed bird a little smaller than a robin, with a black-spotted, white breast, is a thrush (6 to 8½ inches), unless it has a very long tail, and then it's a Brown Thrasher. Five thrushes meet this description, all found in the eastern half of the United States, three of them in the West. Telling them apart is difficult, but the Wood Thrush is easiest to identify, having very big, tear-shaped spots, while the others' spots look more spattered.

AUGUST 16

Obedient Plant

Back on Beaver Island, I discovered a new plant that appeared in my absence, slender spikes of tender, white, snapdragon-lipped flowers. There seemed to be plenty of them, so I picked one. Back home, I identified it as Ladies' Tresses, but I thought it odd that my book didn't mention the square stem on my sample. I hoped it wasn't Ladies' Tresses, as that turned out to be a protected plant in Michigan.

Still puzzled, I put it in a glass of water and before long Mary came over. "That's a *Physostegia virginiana*!" she exclaimed when she saw it. "I haven't seen one in years. You should never have picked it! It's rare!"

Mary is probably right, but I couldn't find it in the only guide I have that tells me which wildflowers are protected in Michigan. The plant was in most of the other books, however. The common name is Obedient Plant: The flowers stay for some time where one positions them. Also called False Dragonhead, its range includes most of the eastern half of the United States.

🐦 🐦 🐦

A SQUARE-STEMMED "SNAPDRAGON"

Many snapdragonlike wildflowers belong to the snapdragon family, but the Obedient Plant is a mint and therefore has a square stem (see page 290), which helps distinguish it from similar wildflowers. The Obedient Plant can flower white, pink or rose, often with freckles on the lower lip; the leaves are narrow, toothed and paired. Turtlehead, with white or pink blossoms that really do look like turtleheads, and white Ladies' Tresses are similar, but the stems of both are round.

AUGUST 17

Highbush Blackberry

The blackberries are getting ripe, and unable to resist such things, especially on a sun-sweet August afternoon, I drove out to the house of some friends and, working my way down their thicket-lined drive, dropped fat berries by the handful into a soup pot. Greedily reaching far into the brambles, I often startled other munchers—chipmunks, birds and mice—bramble fruits (raspberries, dewberries, blackberries, etc.) being a favorite food of North American wildlife. I wasn't worried about an ursine confrontation, as I often am when berry-picking: There are no bears on Beaver Island (no skunks, known Lyme Ticks, or poisonous snakes, either, nor shark fins in the surf). Ripe berries, swollen and glossy, came away with a gentle pull. In twenty minutes I had picked enough for a pie, but by the time I got home, having stopped several times to chat and snack with friends, only enough remained for a sundae.

I've often wondered how I could tell a blackberry or red raspberry cane before it fruited, or before the berries were ripe, so I brought home a sample bramble to compare with a raspberry cane taken from the patch by my cabin. Although both had almost identically shaped, compound leaves, with three to five doubly serrated leaflets in each, seeing the two side by side helped me see obvious differences (see below).

Technically, each blackberry is a group of *drupes*, tiny, individual "berries" with very hard-coated seeds, which tend to get stuck in teeth. "Drupe" does not evoke summer for me, however, so I'm sticking with "berry." There are so many kinds of blackberries all over North America —they hybridize freely—that I can't guarantee that mine was Highbush Blackberry, one of the most common varieties in the Northeast, but since it had leaves that were velvety beneath, it met the description.

‌ᴥ ᴥ ᴥ

RASPBERRY BUSH OR BLACKBERRY?

The fruiting plants are the easiest to identify: Raspberries—both Red and Black varieties—are thimble-shaped when picked, with a hole where they were attached;

Blackberries have no hole. Even if it isn't berry time, you can still tell: Canes covered with thin bristles are likely Red Raspberry (not Black Raspberry); five-sided canes (some Blackberries have round canes), with scattered but serious, blood-drawing thorns, are Blackberry. There are other brambles, of course, but this is a start.

AUGUST 19

Yellow-rumped Warbler

Everyone seems worried about an early, humdinger of a winter coming. Friends who raise foxes have noted unusually dense coats; Don's seen blackbirds flocking; and even south, in Saugatuck, a beekeeping friend is harvesting his honey weeks early because the goldenrod (the nectar of which bees love but which makes the honey sugar over and taste bitter) is already blooming. This morning I began seeing signs, if not of an early winter then, at least, of approaching fall: The black breasts of a large chorus of joyfully chortling starlings were Swiss-dotted winter white; the Cliff Swallows had entirely abandoned their lighthouse mud-jug nests; and I think I identified a Yellow-rumped Warbler in fall plumage.

I really impressed myself with that last one: Identifying a warbler in fall, when its bright mating garb has faded, is a triumph for most birdwatchers. It might even have been an immature warbler, since adolescents outnumber adults this time of year, and the upcoming generation of Yellow-rumped Warblers somewhat resemble their fall-feathered parents. My best clue that I was seeing a Yellow-rumped Warbler was, of course, the yellow rump, but that really wasn't enough: Several other warblers also are yellow-rumped, and one, the Cape May, looks so similar in fall that I'm not completely sure which bird mine was.

ช่ ช่ ช่

THE ONLY NORTHERN WINTER WARBLER

The Yellow-rumped Warbler, although it migrates in huge flocks, is the only warbler that sometimes remains north in winter, being one of the few warblers that can abandon their preferred insect diet for berries, Redcedar, sumac, goldenrod and bird-feeder fare. One of our most abundant warblers, it ranges over most of North America. The Eastern version is white-throated; in the West, the throat is yellow.

AUGUST 21

Northern Red Oak

A stiff wind blew up early this morning, shaking my cabin's canopy of oak boughs. Acorns pelted down on the roof: *Thucka-thucka-thucka-thucka-thuck.* Once begun, the siege was on. I felt as if I were inside a popcorn popper, as acorns rolled off the eaves and plopped off the ground. Risking a direct hit, I went out and picked some up. They were green, about an inch long, with flattish brown caps and stem-end in the middle.

There are so many kinds of oaks, some of which interbreed, that an amateur nature buff, perusing the twenty oak pages in her tree guide, might well be forgiven for panicking and calling an oak an oak. Anyone can go one better than that, however. Oaks do fall into two groups.

Since my leaf had sharp lobes, I checked the Red Oak section of my guidebook, where I found four kinds of oak that looked just like it. Beaver Island is a comforting place to start with oaks, however, because there is essentially only one kind here: Northern Red Oak. None of the similar species—Black Oak and Scarlet Oak—venture this far north. Still, the ranges of these oaks do overlap in other places; for the more courageous, here's a closer look.

❧ ❧ ❧

OAKS: THE RED AND THE WHITE

An oak-shaped leaf with sharp-tipped lobes and/or sharp bristles is usually a kind of Red Oak. Not all oaks have the leaf shape we think of as oak, but the Red Oak section is a good place to begin with any bristle-edged leaf. An oak-shaped leaf with rounded lobes is a kind of White Oak. White Oak acorns mature in one year; Red Oak acorns take two.

THREE LOOK-ALIKE OAKS

All these oaks are found in the eastern half of the United States and have leaves with lobes that curve in about halfway to the middle vein. To tell them apart, look at the leaf, the acorn or the leaf buds at the growing end of the twig. (Note: These tips may not always work, since oaks tend to hybridize.)

LEAF:	Both sides dull and smooth	Northern Red
	Glossy on top, hairs underneath	Black
	Shiny on top, feathery tips	Scarlet
ACORN:	Flattish, beretlike cap	Northern Red
	Bowl-shaped, fisherman's cap	Black or Scarlet
LEAF BUDS:	Smooth (at least partly)	Northern Red
	Hairy	Black
	Often white-tipped	Scarlet

AUGUST 23

Fringed Gentian

I've been out to Gull Harbor three times in the last three days, and each time I've found intensely sky-blue Fringed Gentians among the delicate, ankle-high Brook Lobelia, also sky-blue but with tiny, lipped blossoms. Both wildflowers were thriving in a graveled area where one would think only the sturdy could survive. My visits, however, have not been well-timed—the sun has been either down or behind the clouds—so the Fringed

Gentians have been swirled shut tight as umbrellas. Today I went out at noon, when the sun was hot and high and the air glittered with dragonflies, and I caught the gentians open at last, vaselike whorls of four fringed petals each, flat out and facing the light.

Fringed Gentians are too beautiful for their own good. People pick them, killing future generations. I knew what I saw was a Fringed Gentian from having read, probably thirty years ago, William Jennings Bryan's famous description in his poem of the same name: "Blue—blue—as if the sky let fall/A flower from its cerulean wall." But I didn't know that there were two kinds.

ਦਾ ਦਾ ਦਾ

FRINGE PLUS BLUE EQUALS FRINGED GENTIAN

An intensely blue or violet, vaselike flower with fringed petals is likely one of two Fringed Gentians. Mine may have been the Smaller Fringed Gentian, with very narrow, smooth leaves, growing no higher than eighteen inches, flowers less than an inch across, found in boggy places in the Midwest. The Fringed Gentian proper is twice as tall, with flowers twice as big, leaves twice as wide, found throughout the eastern half of the United States and southern Canada. *Do not pick!*

AUGUST 24

Killdeer

The first time I saw a Killdeer, grubbing in a newly plowed field, I thought I'd found a rare bird, its zebra-striped head and neck were so striking. When it flew, it screamed frantically over and over: *Kill-dee! Kill-dee! Kill-dee!* That was last spring. I've been seeing Killdeer all summer on Beaver Island, too, but I haven't been sure that's what they were, for their call is a rising *Dee!* or a repeated, alarmed *Dee-dee-dee!* Early in

---

PLOVER OR SANDPIPER?

Shorebirds—birds usually found on the water's edge—are myriad and confusing, few more so than plovers and sandpipers, both of which carry compact, short-necked bodies on short, thin legs. To tell which is a plover, look for a short, dovelike bill and a start-and-stop gait.

---

summer, the birds would flap screeching from their ground nests along Gull Harbor when I'd strolled near, although none feigned a broken wing as they are reputed to do when young are threatened. Flying, they moved so fast on narrow, pointed wings that I could never count the neck rings properly. Adult Killdeer have two rings, while other banded plovers have only one.

I finally went out today to count neck rings. It seemed to me that the two birds I saw had only one. They were Killdeer, though: Immature Killdeer have only one neckband. And their legs gave them away.

Killdeer are common near water and around agricultural areas throughout most of North America. In fall, they often fly too high to be seen, but, like Canada Geese, they are vocal in flight, and you'll hear them.

☙ ☙ ☙

BLACK-BANDED SHOREBIRDS

To identify a plover—a robin-sized shorebird—count the neckbands: Two bands around its neck, it's a Killdeer. Listen for its shrill, insistent *Kill-dee!* or *Dee-dee-dee!* or rising *Dee!* The Semipalmated Plover, although resembling the Killdeer, has only one neckband, as does an immature Killdeer. When in doubt, look at the legs: Killdeer legs are flesh-colored; Semipalmated Plover legs are bright orange. That won't work on the Atlantic and Gulf coasts, however, where the single-banded Wilson's Plover has pink legs, too!

AUGUST 25

Spotted Sandpiper

I haven't noticed many sandpipers here, but I saw one yesterday, when I was looking for Killdeer. It was white-breasted and brown-backed, and it bobbed alone along the shore of the Gull Harbor marsh on short, thin legs. The guidebook was little help, showing pages and pages of white-breasted, brown-backed, short-legged sandpipers. If I hadn't noticed the pink legs, I probably wouldn't mention sandpipers at all, but I've found that leg color on shorebirds can sometimes narrow the field considerably. In this case, it narrowed the field to one: The Spotted Sandpiper, even if it didn't have spots. As it turns out, the Spotted Sandpiper is probably one short-legged sandpiper that an amateur can identify pretty surely.

The Spotted Sandpiper is one of the few shorebirds to nest as far south as the United States. In spring, the female Spotted Sandpiper usually arrives first to fight other females for territory. When the males arrive, she'll often mate with one, lay the eggs and leave him to tend to the nest and young while she finds another mate and repeats the procedure. With each female keeping up to four or five males busy on as many nests, it's no wonder that the Spotted Sandpiper is one of the most common sandpipers in North America, found just about everywhere, from wet meadows to ocean beaches.

ᘓ ᘓ ᘓ

ONE EASILY SPOTTED SANDPIPER

If you see a lone, short-legged sandpiper in summer—except when wintering on the southern coasts, the Spotted Sandpipers rarely hang out in crowds—look for big, round, black breast spots, much like those on a Wood Thrush but like no other sandpiper's. In fall and winter, the Spotted Sandpiper loses its spots (immature birds are also "spotless"), so look for a black-tipped, orange bill, pink (although sometimes yellow) legs, and a constantly bobbing tail (from which comes its nickname "Teeter-tail").

AUGUST 27

Common Burdock

We are having a wonderful heat wave; it wouldn't feel like summer without days like this. I spent nearly the whole of yesterday in my swim-suit, sitting for hours in the shallows weaving cedar bark baskets while a breeze blew bulrush whips in my face. Today was hotter yet, but I worked anyway, periodically stepping outside my cabin door to pour water over my head.

On one of these breaks, I took the winding path to Mary's, had a chat and brought home on my socks big inch-wide burrs, wickedly stuck there with tiny, stubborn hooks. The plants I suspected grow along the path, which runs crookedly through a patch of Common Juniper, Smooth Rose and Wild Raspberry bushes. I noticed these right away when I came here in May, clumps of huge, soft, ruffled leaves on strawberry-red stalks that I mistook for rhubarb. As summer went on, the plants grew four or five feet high, big, fat stalks bursting into bushy, feathery bloom. I traded my rhubarb theory for thistle; the purple flower head rays sprayed from a bristly, bulbous base that resembled the burrs I snagged. If the burrs came from this head-high plant, however, what were they doing in my socks?

I retraced my steps and found the answer: Someone had trimmed the huge plants off at the ground, to keep them from blocking the path. New shoots had established themselves, but without all summer to grow, they flowered—and seeded—at calf and ankle level. I finally identified this plant I'd been puzzling about all summer: It was Common Burdock, also known as "Cocklebur."

❧ ❧ ❧

THISTLES AND THISTLELIKE BLOOMS

Several kinds of wildflowers, each with several-to-many variations, feature thistle-like blooms: feathery pink or purple rays sprouting from a hard, bulbous base. You can distinguish Common Burdock and Giant Burdock (a taller version of Common Burdock, with bigger burrs) by the bristle-sharp spikes on the flower bulb. You can also often tell these plant groups apart by the leaves:

THISTLELIKE BLOOMS	LEAVES
Thistles	Edged with painful spikes and thorns
Burdocks	Big, soft, ruffly leaves at base
Knapweeds	Divided, very narrow leaves
Blazing Stars	Single, very narrow leaves

AUGUST 29

Eastern Red-tailed Hawk

I have seen a lot of Red-tailed Hawks this year, but at the time, I wasn't sure what kind of hawk they were. Yesterday morning, however, I watched a Red-tailed Hawk soar circles above the small picnic area next to Barney's Lake, where I was eating a breakfast sandwich and soda.

I'm learning to pick out perching Red-tailed Hawks, especially along the highway. Like Eastern King Birds, they often choose high, exposed places, like dead branches, and also like King Birds, Red-tailed Hawks in the eastern half of the United States are very light underneath and quite dark on top. (There are many color variations of Red-tailed Hawks in the West.) Even if I can't spot the rusty-red tail, a difficult trick on the freeway, I often can pick up the contrast between the wings and breast of a perched Red-tail.

But I've never been able to identify a soaring hawk. Yesterday's hawk, however, gave me a good ten minutes of hawk-gawking. When I really began to look at my hawk guide closely, I realized that identifying a soaring Red-tail was not hard at all.

The Red-tail is fun to look for along highways and rivers. I once counted thirty-four hawks on a three-hour drive across Michigan, and I'm sure most of them were Red-tails, probably the largest common hawk in the United States. Identifying a Red-tailed Hawk in the West becomes more difficult. There the Red-tail has a dark phase that resembles several other hawks common in the West, as well as two other subspecies, all of which may interbreed.

☙ ☙ ☙

HOW TO SPOT A RED-TAILED HAWK

In the East, a large, soaring hawk with a short, bandless, fan-shaped tail is probably a Red-tailed Hawk. (Most fan-tailed hawks have banded tails.) If you can see red on the tail, it's definitely a Red-tail. Look also for a streaky brown belly band over a speckled, cream-colored body. The underwings are light and speckled, and the head appears hooded. For a large, perched hawk, look for a sharp contrast between wings and lower body: The Red-tailed Hawk is the only large, eastern hawk with white feathered legs.

AUGUST 30

Green Darner

Maybe it's the heat wave, but all of a sudden the shoreline, gardens, yards, woods and roads are alive with enormous dragonflies. It's an all-day air show; I even see them in the evening. Bill told me yesterday that on his way home for supper he drove his van through such thick clouds of huge dragonflies, he almost went off the road. He'd never seen anything like it. Until recently, most of the dragonflies around here have been the smaller, shorter-bodied skimmers and other slim, bright types, but now we are being dazzled by one of the biggest, fastest dragonflies of them all: the Green Darner.

An hour ago, I came upon one that had lit on a raspberry bramble. I stopped short just a foot away and watched it stretch out in the late morning sunlight, looking so polished that I was sure it was freshly hatched. It wore a bright green spring jacket, with a matching nose plate; the narrow tail, almost three inches long, glowed bronze with lead-gray rings; large, honey-tan eyes met in the middle like ski goggles; two pairs of transparent wings stretched at least four inches from wing tip to wing tip. When I got home, I identified it easily, the first dragonfly pictured in my insect guide.

🐸 🐸 🐸

BULL'S EYE: A GREEN DARNER

Size is the key: Darners are our largest common dragonflies, and some of the fastest, with wingspans up to four inches and bodies over three. Green Darners have bright green jackets and tails varying from bronze to blue or gray. If you can get close, look for the tiny "bull's-eye" spot where the tan (or light brown) eyes meet. Although it's more common in the East, the Green Darner is found near water all over North America.

September

Buckeye

Several weeks ago a friend and I kayaked to a quiet bay, where we took cover from a drizzling rain under thick cedar boughs behind the sandy beach. There, next to a tiny creek, we found watercress, bright red mushrooms and a translucently stemmed plant called Jewelweed, which wasn't yet in bloom. Today, thinking the Jewelweed might be flowering, I headed back to find it.

This time I went on foot, a two-mile trek along a stony beach. Before I'd gone far, I spotted a butterfly clinging to a scrap of wood on the warm sand, slowly opening and closing its inch-wide wings to the sun. As I crouched over it, an oddly human-looking "eye" seemed to stare at me from the upper outside wing. I knew it wasn't a real eye, but it was unsettling, and when the wings opened, I saw six more eyes, these filled with brilliant tropical colors, ringed with white and black, resting like jewels on feathery velvet brown.

I didn't understand at the time why the butterfly tolerated my presence for so long, but I later learned that it was probably due to the nippy nature of the day. Butterflies can't fly until their slim bodies reach a temperature of at least eighty-one degrees. Lacking the internal heating system of birds and mammals, they need to sunbathe, and the cooler the day, the more time they spend doing it. I was able to stare at the butterfly long enough to memorize the details, and once home, I identified it easily. The only species of its genus to venture out of the tropics, it was a Buckeye, named for the eerie outer "eye."

Although found throughout most of the United States, Buckeyes are more common in the warmer, southern states. In fall, clouds of Buckeyes from the North are often seen, especially in the East, drifting south.

&a &a &a

IF "EYED" BY A BUTTERFLY...

A fawn-brown, medium-sized butterfly (smaller than a Monarch, bigger than a Cabbage) that appears to be staring at you from closed wings is likely a Buckeye. There are other brown butterflies with eyespots, but none like this: The Buckeye's round "eye" is set on an oval of white, and brought alive by the fleck of white in the "pupil." Look also for two orange bars on the upper wing. Inside, find six eyespots, three on each side, brilliant pink and blue.

SEPTEMBER 2

Boneset

When two plant habitats meet, as a woods on a meadow, interesting things can happen, and yesterday's jaunt around the other side of the harbor was proof. To my left, sailboats swooped out of the harbor over glossy green swells; to my right loomed a dark, damp cedar woods. Between these, along the narrow strip of watery, mushy beach, I found nine or ten plants I had never identified. I thought it odd that six of these were red-stemmed: low-growing Sand Cherry with its speckled, sour-smelling bark typical of cherry; Wormwood, red wooly stems spiraled with wands of tiny, podlike green "flowers"; puffy-pink clusters of Joe Pye Weed; low-growing Huron Tansy, with reddish, fernlike leaves smelling strongly medicinal; Silverweed, spreading like strawberry plants on bright red runners; and oddest of them all, a plant called Boneset.

None of my guides mentioned red stems for Boneset, or purple leaves, either, which were present on my find, but there was no question about its identity. Boneset's peculiar leaves look to me like flying, lanky-winged birds.

Boneset, also known as Thoroughwort, is common in moist clearings, ranging throughout the eastern half of the United States and most of Canada.

❧ ❧ ❧

THE "PIERCED LEAF" PLANT

Each pair of pointed, triangular Boneset leaves is joined at the base, forming a long, thin diamond that appears to be pierced by the plant's stem. Each diamond grows crosswise from the one 2 inches or so beneath it. If you see a knee- to shoulder-high, bushy plant with flattish clusters of tiny, grayish-white tufts, look for these odd wrinkled leaves. Just about everything on Boneset is faintly fuzzy: the stem, the undersides of the leaves and the flowers with their threadlike "petals."

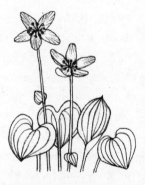

SEPTEMBER 3

Grass of Parnassus

The bay hike continued: When I reached the bay I was looking for the other day, I discovered several plant environments. Along the lake lay the sun-baked, sandy dune, scattered with grasses and many of those red-stemmed plants; the backdrop was dense cedar; and along the outside edge of the woods ran a thin strip of moist semishade where thrived a lovely foot-high flower bed featuring a patriotic combination of blue Fringed Gentians, some bright red seed clusters I couldn't identify and startling white flowers called Grass of Parnassus.

The name seems something of a mystery: Grass of Parnassus is neither a grass nor found on Parnassus, or anywhere else in Greece. Grass of Parnassus grows over most of eastern North America and the southern half of Canada. Although there are other similarly sized, starlike white flowers in the same range, Grass of Parnassus has a couple of distinctive traits.

᭰ ᭰ ᭰

THE PIN-STRIPED STAR

A nickel- to quarter-sized five-petaled white flower with prominent green veins is likely Grass of Parnassus, especially if there is just one round leaf clasping the long, thin stem. The rest of its spade-shaped leaves are found at the base. Reaching one to five feet, Grass of Parnassus blooms in late summer and fall. Two similarly starlike, white flowers are the six-petaled Wood Anemone and the seven-petaled Starflower, but their petals are not green-veined and both are under one-foot tall.

SEPTEMBER 4

Jewelweed

When I reached the fragrant, dark woods, my feet sank ankle-deep in damp, brown cedar lace. I fought my way through thickets of fallen branches, climbed over rotting logs, pushed through adolescent trees, and finally reached a tiny brook. A decayed log bridge built only a foot over it snapped under my weight soaking one foot. When I finally sat down, exhausted, there it was, the plant I'd hoped to see in bloom: Jewelweed, hung with delicate, inch-long, nasturtium-orange blossoms. I hadn't been sure I would be able to find it again.

Jewelweed is named for its pendant flowers, which hang like gemmed earrings on long stems. The horn of plenty–shaped blossoms are easy to spot, for, with the exception of hawkweed and lilies, orange is not a common wildflower color. The leaves are unusual, too: When submerged, the undersides turn a gleaming, mercurial silver; late in fall, the seed pods, when touched, explode, shooting seeds as far as four feet! There's another reason to know Jewelweed: The leaves give quick relief from a Poison Ivy rash, and the two plants often grow in the same conditions.

ᵰ ᵰ ᵰ

JEWELWEED: ORANGE IN THE SHADE

A tall plant hung with small, orange, horn-shaped flowers (a northern variety has yellow flowers) is probably Jewelweed, especially if limp, spear-shaped leaves with rounded teeth are found on the translucent stems. Jewelweed is a kind of impatiens and, wilting easily, thrives in wet, shady places. Although my find was only two feet high, thickets of Jewelweed can reach eight feet. Found throughout North America, Jewelweed may bloom from June through September.

SEPTEMBER 6

Jumping Spiders

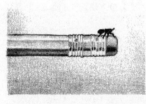

Yesterday afternoon I was stretched out on my bed—a mattress on the floor—taking notes on Jewelweed, when a half-inch black spider charged over the front edge of the mattress and then continued across the blue-lined pages of my spiral writing pad. I gently nudged it back on the floor with my ballpoint pen. Seconds later, the spider was back again, traversing my notes.

Now, I've been sharing my cabin with several hundred spiders this summer, and I've found them to be quite retiring. They've got their nooks, I've got mine, and we pretty much leave each other alone. If I brush a spider off my sheets, it gets the hint. This one, however, didn't. It returned a third time. I held my pen in front of it. It hopped on and, much as a ladybug might, crawled toward my hand. When I shook it off, it hopped back on. When I raised my pen high above the notebook, the spider leaped off, trailing a silken lifeline. Landing, it reared back and threatened me with a pair of crablike front legs.

Although I couldn't find this particular spider in my spider guide, its entertaining behavior made it without doubt a jumping spider. At half an inch, however, it was bigger than most jumping spiders. I'd never heard of jumping spiders before, so I was happy to learn that for all their aggressive behavior, they probably pose less threat to humans than any other spider group.

&a &a &a

THE "PET" OF THE SPIDER WORLD

A tiny, tanklike spider that jumps a little like a flea when nudged is probably a jumping spider. Jumping spiders, found throughout North America, can be pleasantly sociable and fun to watch. You'll have to look hard, though: Most jumping spiders grow, excluding legs, no larger than one-fifth of an inch, although some reach a whopping half inch!

SEPTEMBER 8

Pink-spotted Hawk Moth

Last night, watching the sun set behind great pink piles of clouds over the harbor, I noticed what I thought were two hummingbirds hovering in a patch of Soapwort. I'd never seen hummingbirds in the evening before. Approaching them slowly, I realized I was seeing, not hummingbirds, but two enormous moths, each with a hummingbird's wingspan (five inches), hovering with blurred wings while long, threadlike tongues probed each flower for nectar. They flew within inches of my pant legs while I crouched over their busy beelike bodies. Bright pink-and-black stripes, intersected by a thick middle line, ringed their abdomens; the hind wings shimmered a dusty rose; long, narrow top wings blurred gray. The bodies were about half as long as the wingspan, unlike the Ruby-throated Hummingbird's body, which is slightly longer than its wingspan.

Once home, I easily identified them as Pink-spotted Hawk Moths, but I was lucky: Out of about thirty-five ring-bodied hawk moths, also called sphinx moths, pictured in my moth guide, only one had pink rings instead of yellow. Although identifying a particular species of hawk moth is difficult, especially when the wings are a blur, hawk moths as a group have so many distinctive features that they are easy to pick out.

🦋 🦋 🦋

THE RING-BODIED HAWK MOTH

A moth with a fat body and brightly ringed abdomen—usually yellow and black, but sometimes pink and black—is likely one of the many kinds of hawk moths. Look for unusually long and narrow wings and a sharply pointed tail end. A hawk moth hovers like a hummingbird, sipping nectar with an extremely long, uncurled tongue. Some species can be seen at dusk or dawn. Hawk moths—with wingspans varying from one to seven inches—can be found just about anywhere in North America.

SEPTEMBER 10

Canada Goldenrod

On the evening before my departure from Beaver Island, the Shamrock put on a clamfest, a happy coincidence. Mary, Glen, Bill and I feasted on lobsters, grilled chicken and clams and danced past midnight to a live band. Early the next morning, my loyal friends roused themselves to wave me good-bye on the eight-thirty ferry. I waved back, high above them, wedged in the point of the prow, watching my three friends shrink to dots, then, through my binoculars, seeing them turn back to their lives. Below, my van was loaded, all spare spaces stuffed with strips of cedar bark I'd peeled and hung to dry on my clothesline between the hand-washed socks and underwear (there was no island laundromat).

The voyage took nearly three hours. Waves crashed over the top of the big boat. We lunged side to side and front to back, somehow simultaneously. We'd take off on a swell, then sink like a stone. Water poured down the walkways outside the lounge windows. The supply of barf bags was going fast. Pale passengers lurched about the boat, unable to take one step without being thrown hard against something. I sat tight. I didn't get seasick, but I felt cold. I visualized the interior of my van below as a sort of blender, the contents of my life being processed into a smooth, homogeneous mass something like Silly Putty.

To survive, I thought about what I had left. I had given myself something I hadn't known since childhood: a real summer. I had gone swimming often. I recalled the refreshing shock of cool water after sweating all afternoon at my computer; and later, the languid satisfaction induced by bath-warm waves; and, best of all, the delicious thrill, on a hot moonlit night, of leaving my swimsuit wadded on the end of the dock, and slipping into the silky black lake. I'd learned to dance at the Shamrock; I'd never danced much before. One night I danced so hard that a friend whose family tree, like mine, is rooted in the Netherlands told me, when he could stop laughing, that in all of his considerable life he'd never seen a person of Dutch descent move as fast as that. I'd taken naps whenever I felt like it. I'd made lots of baskets, falling in love with cedar bark, its rose and lavender hues. Wet, it felt like leather. Sometimes it came off a log as pink and white as a newborn baby, perfumed, damp and pliable.

Thoughts like these got me to Charlevoix, where I began the four-hour drive to Douglas. My van was fine; I had packed it so tightly that nothing had shifted. The trip was fine, too. The roadsides were glorious with waist-high swaths of goldenrod, which made me smile: Once long ago, when I brought my mother, famed for her cool in a crisis, a big bouquet of goldenrod, a plateful of spiders could not have gotten half such a rise out of her.

My allergy-prone parent was suffering from the common delusion that goldenrod causes hay fever. Actually, ragweed and its relatives, not goldenrod, is responsible for most allergic reactions to pollen. Goldenrod is pollinated by insects, its brilliant color offering plenty of sex appeal, but ragweed flowers, although also blooming on "rods," are so dully colored and scentless that bugs and birds ignore them. To get the job done, ragweed must cast its pollen—huge amounts of it to make up for the inefficiency of the method—upon the winds.

My friend Ellen once attempted to analyze goldenrod's unusually intense color. "If you are an artist, you are aware of how very different one yellow can be from another," she told me, waving at her colorful, tiered garden. "Goldenrod doesn't fall easily into any particular yellow. It isn't ochre and it isn't chrome. It really does look golden."

It was Canada Goldenrod I saw along the road, tall and blooming in rich, intensely colored plumes. But there are many kinds of goldenrod, many quite similar to one another.

ම ම ම

WHAT KIND OF GOLDENROD?

A richly golden, plumed fall flower cluster is likely some kind of goldenrod. The problem is: Which kind? There are sixty to a hundred goldenrods, depending on which book you read, found in North America. Goldenrod blooms in clusters of tiny yellow, rayed blossoms, but the arrangement of these can differ widely. Peterson's *A Field Guide to Wildflowers*, the best help I have found on the subject, devotes thirteen pages to goldenrods, dividing them into these types:

1. Plumelike, graceful clusters
2. Graceful, elm-branched plumes
3. A clublike, showy cluster topping each stem
4. Small clusters blooming along a wandlike stem
5. A flat-topped cluster, much like a bright yellow Queen Anne's Lace

The Canada Goldenrod, common along midwestern roadways, fits the first category, one of the most easily recognized types of goldenrod. The Bluestem Goldenrod in my yard fits the fourth.

SEPTEMBER 12

Asters

Blooming among the goldenrod along yesterday's roads were clouds of asters: billowing, pale, lavender blossoms that looked from the car like lilacs on long stalks. The lavender and yellow, opposites on the color chart, created a striking display against the lush late-summer green. Also called "star flowers," asters are flower heads, built daisy fashion with yellow middles and ray flowers. Often, however, the middles darken to purple or dark red as the individual flowers that compose them are pollinated.

Aster petals are finer and generally more numerous than daisy petals, and the flowers are usually clustered. Despite wide species differences in size—blooms can vary from very small to two inches across—and color—anything from white, yellow, lavender or blue to shades of pink from pale to deep rose—recognizing a member of the aster family is surprisingly easy. Few other wildflowers closely resemble its structure.

Asters can bloom quite late into the fall and are often some of the last blooms seen before winter. The nondescript White Aster (also known as the Snow Aster) is loved by beekeepers as a nectar source for their bees' winter store of honey, after the summer's honeycombs have been harvested.

(See also page 263.)

 ᐜ ᐜ ᐜ

ASTER'S EASY; WHAT KIND OF ASTER'S NOT

If it looks like an aster, it probably *is* an aster. The wildflower I confuse with it is fleabane, a small white or pink asterlike flower with a fluffy fringe of fine petals finer than aster petals—too many to comfortably count. Fleabanes generally bloom from early summer to fall, however, while most asters don't begin blooming until late summer, so a spring "aster" is likely a fleabane. (See page 207.)

Telling aster species apart is another matter. There are reportedly about 250 species of wild asters in the world, over 70 in the United States, and they happily interbreed. A guidebook usually helps with the most common varieties. Sometimes, at least for me, nothing helps!

SEPTEMBER 13

Ovenbird

I'm moving back into my lakeshore apartment now, and although the inside is cluttered with boxes, the outside is lush with late summer. An Ovenbird sang in the maple outside my bedroom this morning, a loud, rapid *Teacher teacher teacher* (the Ovenbird is sometimes called the "teacher bird"). I knew it not just from its distinctive song, but because last spring, below a picture window, I'd found one dead. I brought it inside on a small garden spade, laid it on my table and drew it, which forced me to really look at its small sleek shape, olive top, darkly spotted white breast, bright yellow crown and flamingo-pink legs. Never having seen anything like it before, I thought I'd found something rare. (Who's ever heard of an Ovenbird?) It turned out, however, to be one of the most common wood warblers around, but due to its size and leaflike wings, it's more often heard than seen.

The Ovenbird is a ground-lover. Unlike a hopping robin, it walks over fallen leaves looking for insects. Its tiny Dutch-oven nest, from which comes its name, is built near the ground, too.

🐦 🐦 🐦

HOW TO TELL AN OVENBIRD FROM A WATERTHRUSH

Look for bright pink legs and no white eye streak. The Northern Waterthrush (found in the northern part of the Ovenbird's range) and the Louisiana Waterthrush (found in the southern part), are similar in size, shape and coloring to the Ovenbird, but they have paler legs, white eye streaks and sing different songs. The Ovenbird ranges throughout most of North America, except the most western quarter.

SEPTEMBER 14

Common Ragweed

I don't suffer from hay fever, so I've never much cared what ragweed, an unattractive plant at best, looked like. People with pollen allergies, however, would probably do well to know ragweed when they see it. Common Ragweed and Giant Ragweed are such prolific pollen spreaders that some states have made strong efforts to control their growth. In Michigan, for example, it is illegal to have mature ragweed on private property after the middle of July.

Ragweed's sin against the human sinus is due partly to its habit of pollination and partly to the barbarous structure of the pollen itself. First, ragweed is wind, not insect, pollinated. Ragweed's long strands are laden with tiny green flowers full of yellow pollen, which is exploded into the air, an adaptation which Neltje Blanchan, author of a wonderful old commentary on wildflowers, *Nature's Garden* (Grosset & Dunlap, 1900), calls "extravagant." It's far more efficient and elegant, she points out, to get help from the birds and the bees. Unfortunately, ragweed hasn't got the scent or color to attract them. What makes ragweed even more noxious, however, is the structure of the pollen, which is released from male flower heads: Each pollen grain bristles with microscopic barbed hooks that cling to anything they touch, including human tissue.

I went out looking for ragweed and came back with a number of possibilities. Wormwood, which can also aggravate sinuses, can closely resemble ragweed, some species having deeply lobed leaves and green flowers, but that's just as well: Allergy sufferers will want to stay clear of Wormwood, too.

❧   ❧   ❧

A TRICK FOR TELLING RAGWEED

Common Ragweed flower stalks are long and substantial, the leaves many-lobed. The plant grows to about three feet and prefers dry places. Great Ragweed likes wet places, can grow two to fifteen feet and has similar flower stalks, but large, goosefoot-shaped leaves. To identify either ragweed, look for long plumes of small, greenish flowers without leaves between the flowers, unlike similar plants, like Lamb's Quarters, Wormwood and Mugwort, which have small leaves among the blossoms.

SEPTEMBER 15

Tulip Tree

I think autumn is a good time to take on the trees. Trees often give more clues in the fall than in other seasons. Each tree turns a particular color, and with a little luck the leaves, often frustratingly out of reach, float right into my hand. I'm not sure why naming the leafy trees makes me so apprehensive. Perhaps it's their tremendous size, or because to me one mass of green, so far above my head, looks much like another. So I'll make it easy on myself. I'll begin with the Tulip Tree.

The Tulip Tree is my favorite tree on my street: straight, tall, with tulip-shaped leaves and cup-sized, greenish yellow flowers. Its leaves are unmistakable, even to a rank amateur: They really are tulip-shaped, and only slightly smaller than a maple leaf. The Tulip Tree is also known as a Yellow Poplar, because the leaves tremble in the wind as poplar leaves do, but it's not a true poplar. It belongs to the magnolia family.

ॐ ॐ ॐ

THREE EASY TREES

These three common trees have leaves so strangely shaped that identifying them is no trick at all. Both the Tulip Tree and the Sassafras are common throughout the eastern United States, while the Ginko, although not a native U.S. tree, is frequently found planted in cities, thanks to its extraordinary tolerance of pollution.

TULIP TREE:	Large, smooth-edged, tulip-shaped leaves
SASSAFRAS:	Large, smooth-edged leaves in three odd shapes: egg, mitten and goosefoot
GINKO:	Smaller, fan-shaped leaves

SEPTEMBER 16

House Wren

I came upon an enormous brush pile on a lumber road today, comprised of the exposed roots of two fallen trees and some discarded brush. What attracted my attention was the great deal of activity inside it. Sparrows, I thought, catching glimpses of flitting, small, brown birds, but there was something odd about the loud whirrs of wings, the dartings. The birds moved so quickly, I had difficulty catching one in my binoculars, but when I did, I recognized a wren from its straight-up-in-the-air tail and curved, slim beak. Knowing a wren is easy, but telling the different wrens apart is a bird of another feather. My wren had no bright white eyebrow, which narrowed the choice to three. I decided it was probably a House Wren, which, according to my guides, often frequents brush piles. Being the most common, it was my best bet.

I was at first confused by the dark coloring on the House Wrens I saw, until I learned that House Wrens darken considerably in the fall. They can be quite sociable, often enjoying human habitations and nesting in all sorts of strange containers. Some people love them for their burbling, happy songs; others dislike their aggressive behavior, for, despite their tiny size, they may waste no time chasing off other birds.

ə ə ə

WREN "DE-TAILS"

A small, energetic ruddy-brown bird with a tail frequently held straight up in the air is likely some kind of wren. Wrens fall into two groups: those with a white eyebrow and those with a barely discernible eyebrow. Of the hardly-any-eyebrow group, the four-to-five-inch House Wren is the most common, found all over the United States and most of Canada. Much of this territory is shared by the smaller, less common Winter Wren, distinguished by the stubbiness of its stuck-up tail. In the East, the uncommon Sedge Wren has a darkly streaked back. In the West, the Canyon Wren is identified by its white throat and upper breast.

SEPTEMBER 17

Bittersweet Nightshade

Yesterday's brush pile was draped with scarlet berry bunches belonging to a beautiful, danger-ous vine common all over North America: Bit-tersweet Nightshade. I'd learned to recognize the small but dramatic shooting-star, violet flowers, but Bittersweet Nightshade is most easily found in fall. It's an odd vine: Having no tendrils and seldom even attempting to twine, it usually sprawls over, under or behind the shrubbery, making it hard to see. In fall, however, the bright red berries pop up all over the place. There's rarely any question about what the plant is, even though there are other small red berries around now: The leaves are so strange that I can pick out Bittersweet Nightshade without flowers or berries.

Bittersweet Nightshade, once known as "Deadly Nightshade," gets its name from its berries, which, though poisonous when green, are eagerly devoured by over thirty kinds of birds when ripe, although not recom-mended for human consumption. (Birds eat Poison Ivy berries, too!) Night-shade was once known as a powerful toxin, so it's no wonder that the tomato, a member of the nightshade family, was once considered inedible.

a a a

PURPLE SHOOTING STARS

A smooth vine with loose clusters of small (half-inch), violet (sometimes white) shooting-star–shaped flowers is probably Bittersweet Nightshade. The only other plant with violet-to-white shooting stars is Horse Nettle, a prickly, erect plant. If you're not sure, look at the oddly shaped leaves. Most nightshade leaves are egg-shaped, with two small lobes at the bottom.

SEPTEMBER 19

Barred Owl

This morning a huge, dark owl flew in front of my car just as I was about to leave my driveway. It flew past at windshield level and, with a silent sweep of wings, vanished into the woods behind the house across the road. I thought it was a Great Horned Owl, but I didn't get a very good look at its head.

I've had three brushes with owls this year. There was the Great Horned Owl I saw last spring (see page 73). Once I found an owl feather next to a live owl's cage in a nature center. The owl had been injured and the center had gone through much red tape to get a license from the federal government to house it.* The feather was generously wide, with slashes of deep brown on tan, and it felt velvety, like the wing of a large brown moth. Each filament was locked to the next, making a soft, delicately ribbed fabric that, if split, was easily repaired. I'd never seen or felt a feather like it before.

My first encounter with an owl was last winter in the middle of a moonlit night, when I was awakened by a rabbit's nightmarish scream close

---

* It is illegal in the United States to handle any raptor—hawk, eagle, or owl—without a license or special permission.

to my bedroom window. It was a shriek of death, almost human. Although I never saw it, I was sure a rabbit had been taken by a Great Horned Owl, the only large owl I knew about, which is why I thought the owl I saw this morning was one. As it turns out, however, it also could have been a Barred Owl, the other big brown owl in these parts (see page 74.)

SEPTEMBER 20

Horsebrier

I was toughing it out through some pathless woods today, getting caught in all sorts of brambles, mostly blackberry, when I got stuck in a very different sort of prickly vine, one with hand-sized, smooth, almost heart-shaped leaves. It was the veins in the leaves that caught my attention: Three big veins curved from stem to leaf tip with two short veins at the sides, and on the underside of the leaf, little prickles ran up each vein.

I brought home a sample and found I'd entangled myself in one of the many greenbriers, vines common in the East along the edges of roads, woods and rivers. Horsebrier has a bigger, wider leaf than other greenbriers, and more widely spaced thorns, none of which occur at the leaf joints, but according to my guides, it's easy to confuse greenbriers. It seems to me to be enough to know a member of the greenbrier group when I see one.

🙰 🙰 🙰

THE THORN-AND-TENDRIL VINE

A wild-growing vine in the eastern United States that has both thorns and tendrils is probably some kind of greenbrier. You can also tell a greenbrier by the veins in the leaves, which start at the stem and travel upward to the tip of the leaf. There are many kinds of greenbrier, the most common of which are Catbrier, Horsebrier, Bristly Greenbrier, Red-berried Greenbrier and Laurel-leafed Greenbrier.

SEPTEMBER 21

New England Aster

This afternoon, at forty-five miles per hour, I suddenly recognized New England Asters. Although I'd never really seen them before, they were everything I'd read about: big, showy, dramatic, numerous and magnificently magenta. New England Aster flowers are big as daisies, and, as with other asters, their yellow centers turn purple with pollination. They can be other colors, too: blue, lavender, purple, white or pink. The leaves and stems feel surprisingly sandpapery. I don't know why they're called New England Asters: They bloom all over the eastern half of the United States and in many southwestern states as well, and they're not the least bit conservative. New England Asters flower later here than many of the smaller-flowered asters do, but now I'm seeing them everywhere. (See also page 263.)

SEPTEMBER 22

Maples

Last week, I wailed about the complexity of maples to Earl, a state park naturalist near here, and Earl said, "Hey, maples aren't so bad. They're really quite well behaved compared to oaks."

Maybe so, but I've been staring at maples for two weeks. Sure, anybody can tell a maple—the big, lobey leaf is pretty distinctive, and the maple leaves always grow in pairs along the stem—but not all maples are alike. This morning, at long last, when I looked up at the leaves spread against the sky by the many huge maples on my street, I began to *see* the leaves. They seemed to fall into three groups for me, depending on what sort of edge they had. Some maple leaves had smooth edges, others had big teeth, and a third group were serrated.

ॐ ॐ ॐ

MAPLE LEAF EDGES AND WHERE TO SEE THEM

If you first narrow the possibilities to those maples growing in your area, and then look at the leaf margins, or edges, identifying a maple becomes less intimidating. Those of us who live in the East have the hardest time identifying maples. Westerners have it easy, although eastern maples are often planted there as ornamental yard trees.

Northeast:	Sugar and Black (Smooth edges)
	Silver (Large-toothed edge)
	Red, Striped and Mountain (Serrated edges)
Southeast:	Chalk and Florida (Smooth)
	Silver (Large-toothed)
	Red (Serrated)
West:	Bigtooth (Smooth) and Rocky Mountain (Serrated)
West Coast:	Bigleaf (Smooth) and Vine (Serrated)

SEPTEMBER 23

Sugar Maple

If trees could move, they'd probably scare me to death. Trees are very large, usually with unreachable leaves, and they are terribly, terribly green. A wildflower feels gentler, attains a predictable shape and categorizes itself immediately by color. But the Sugar Maple seems cozily familiar; it even made the Canadian flag! It also happens to be the most common maple in my neighborhood.

The Sugar Maple leaf has the big, classic shape I have always thought of as "maple": smooth-edged with five obvious lobes. The sides of the top lobe are straight, running parallel to one another. Now I can spot leaves with this boxy top lobe from the ground, even if they are far out of reach.

ঌ ঌ ঌ

SUGAR MAPLE LOOK-ALIKES

In the northeastern quarter of the country, there are two maples that resemble the Sugar Maple. One way to tell the three apart is to look at the leaf underside:

MAPLE	LEAF UNDERSIDE
Sugar Maple	Light and dull, sometimes slightly fuzzy
Norway Maple	Bright and glossy; leaf is broader than it is tall; the leaf stem exudes a milky sap when broken
Black Maple	Light and fuzzy; Black Maple leaves often have only three lobes, with simpler lines

In Florida, it's hard to tell the Florida Maple and the Chalk Maple by the leaves: Both have fuzzy undersides and resemble the Sugar Maple leaf shape. Instead, look at the trunks: The Chalk Maple trunk is fairly smooth and chalky white; Florida Maple bark is ridged and gray.

SEPTEMBER 25

Silver Maple

After struggling with Sugar and Striped Maples, it's a relief to realize that there is one maple that doesn't closely resemble any other: the Silver Maple. A Silver Maple leaf is deeply five-lobed and ornate, with toothy edges. The top (middle) lobe begins at a sort of cinched "waist" and, after three successively narrowing extensions that resemble a cross-legged elf, ends in a jaunty, pointed cap. The back of the Silver Maple leaf is almost silvery. When I look up at the boughs of a Silver Maple tree, I see a moving mass of silver linings, startlingly light next to a more predictable green. Silver Maples are so beautiful that they are often used as street trees in cities and towns, or as ornamental landscape trees.

SEPTEMBER 26

Sulphur Butterflies

Strong offshore winds have been blowing against my windows for two days, and I've been worried about the safety of the millions of birds and butterflies migrating south along the lakeshore. This morning on the beach my anxieties were confirmed: Thousands of drowned and nearly drowned Common Sulphur Butterflies lay at the edge of the surf, their soaked wings faded from sulphur yellow to a pale luna green and weighted down with sand. Some of them were still alive, struggling against the sand that held down their wings like rocks on a sail. I picked some of these up, brushed them off and set them up farther on the beach, hoping that when the sun came out, perhaps a few would dry out and fly away.

Living on a migration path has sensitized me to the danger to birds and butterflies this time of year. Night is a popular journeying time, safer than daylight. I've been told that if I watch through my binoculars at night now, I might see throngs of birds flying across the bright face of the moon. On very stormy nights, I sometimes wake up worrying about all those wings beating through wind and darkness over the heaving, black lake. Monarchs are not the only butterflies to migrate. Vast numbers of sulphurs, Red Admirals, Painted Ladies and others can take on quite lengthy flights.

It's often difficult to identify the species of sulphur butterflies, because there are so many of them, ranging from yellow to green to orange, some with black wing borders, some with spots, but all with a wingspan measuring about an inch and a half, and all about the same shape. They are common in meadows, roadways and parks all over the United States. Watch for their vertically spiraling flights: A male and a female often fly around each other,

---

BUTTERFLY OR MOTH?

Most people look at the bodies: Moths usually have fat bodies, while butterflies usually have thin ones. While this is often true, a whole family of butterflies called Skippers have fat, mothlike bodies. You may have assumed that moths fly at night while butterflies are diurnal, which is usually true, but not always. It's really the antennae that distinguishes a butterfly from a moth: Butterfly antennae end in a club, or swelling, while moth antennae are either feathered, or narrow to a point, like a hair.

---

higher and higher, until the male gives up and falls like a dead weight, sometimes sixty feet toward the ground, while the female drifts down more slowly. This is thought to be the female's way of discouraging unwanted male attention after mating.

SEPTEMBER 27

Sandhill Crane

Yesterday, in pristine autumn weather, two friends and I took a canoe up the Kalamazoo River. After a picnic, we paddled back through the wetlands, rousing flocks of migrating ducks, six or seven Blue Herons, several Belted King-fishers, lots of Killdeer and two Red-tailed Hawks, one of which demonstrated, as if for the next take, a spectacular dive into the drink. The most amazing spectacle of all, however, was the grand finale, also occurring as if on cue: Two elegant lavender-gray Sandhill Cranes, with bright red topknots and wings widespread, danced at the edge of the water, floating into the air and facing off while another stood by and watched.

I had never seen a Sandhill Crane before; I thought Sandhill Cranes performed only for major networks. Once almost extinct, the Sandhill have returned in such strength that they are reportedly becoming a common sight in Michigan. I was surprised at their size: a six- or seven-foot wingspread. As for the dancing, the males do face off, but a mating couple, often faithful for years, will also dance with each other at dusk and dawn, a ritual I might adopt myself, should appropriate circumstances prevail.

ᔰ ᔰ ᔰ

SANDHILL CRANE OR GREAT BLUE HERON?

The Sandhill Crane and the Great Blue Heron are similar in size, shape and color. To tell them apart when they're airborne, look at the neck: The Great Blue Heron sets its neck in an S-curve in flight, while the Sandhill Crane stretches its neck out straight in front, with legs stretched out back, earning it the nickname "Flying Cross." On the ground, the Sandhill's red crown gives it away.

SEPTEMBER 29

American Beech

I've been asked why I don't use the shape of the tree or the trunk, which come more easily to hand, instead of the leaves to identify a tree. The answer is that trees are often shaped as much by available light and the proximity of neighbors as by natural inclination. As for the trunks, most of them look alike to me and, unlike leaves, I can't get them in through the door for a leisurely, centrally heated inspection. The bark of some trees, however, like that of the beech tree, is distinctive.

Beech trees, common along the sandy Lake Michigan shore and found all over the eastern half of the United States, are wrapped in bark like elephant hide: thin, smooth and a light, almost bluish gray, with curved scars. The leaves are oval, about three to five inches long, with widely spaced teeth along the edge and prominent veins, like the skeleton of a fish. They closely resemble elm leaves, but the American Beech's trunk identifies it immediately.

≥⚬ ≥⚬ ≥⚬

TREES WITH DISTINCTIVE TRUNKS

The American Beech is among several trees that are easiest to identify from their bark:

AMERICAN BEECH:	Smooth, slightly mottled "elephant hide"
BLUE BEECH (HORNBEAM; IRONWOOD):	Thin, tight, blue-gray bark with musclelike, sinewy, vertical ridges
PAPER BIRCH:	White bark that peels in long strips
YELLOW BIRCH:	Yellow-gray, shaggy bark that curls like wood shavings
STRIPED MAPLE:	Thin, greenish bark with vertical white stripes

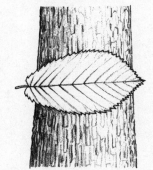

October

Stinkpot

The surf's been up on the beach for days now, delivering butterflies and gull feathers, but the big surprise was a blackish, oval, high-domed turtle that looked like a soap-on-a-rope. I carried it home, thumb and third finger positioned, I found out later, on each of its stinky musk glands. The turtle kept its head and extremities pulled tightly in until it was submerged in a cereal bowl of water on my dining-room table. Slowly, the head emerged, revealing two yellow lines running above and below each tiny, black, glaring eye. These distinctive markings greatly eased identification: It was a kind of musk turtle—fairly common, as turtles go, throughout the eastern United States. Mine had to be a Stinkpot, since the other three North American musk turtles confine themselves to the Southeast.

It is usually best to return a creature where one found it, but I'd never seen a turtle on the exposed beach before. I called Earl Wolf, nearby naturalist, who said he had rarely seen one there, either. "But turtles sometimes get caught in the river current and get swept right out in the lake," he said. He suggested I release the Stinkpot into nearby wetlands. The Stinkpot remained docile, so I set it, bowl and all, on the floor of my van, on the passenger side. By the time I'd reached the Kalamazoo River— about five minutes later—the turtle was gone. I couldn't believe it. I'd glanced at it seconds before!

I pulled the car over, got out and carefully opened the passenger door. No turtle fell out. I searched under the seats, under the foam mattress in the back, in my camping gear. At last, patting the carpet, I discovered a damp trail leading up under the dashboard, and my heart sank. A Stinkpot stuck in the dashboard of my car was not unlike a skunk in the trunk: Not only can a Stinkpot secrete nasty-smelling musk, but it will bite if it can. Much like the Snapping Turtle's, a Stinkpot's shell cannot close tightly, leaving it quite vulnerable.

I stuck my hand up under the dash, reminding myself that turtles have

---

TURTLE SHELL PATTERNS ARE ALL NEARLY THE SAME

You usually can't identify a turtle by any distinctive shell pattern; with the obvious exception of smooth-backed turtles, the *scutes* that make up the *carapace* tend to be the same shape and number on any turtle. There can be wide differences in the *plastron*, but turtles, quick to slip away, usually elude you before you can flip them.

---

no teeth, when out came the Stinkpot through another opening. Musk turtles, it turns out, are one of the few turtles that are good at climbing. ("If a turtle ever falls on your head, or in your canoe," I was amused to read later in Roger Conant's *Peterson Field Guide to Reptiles and Amphibians*, "it will probably be this or one of the other musk turtles.") Holding the writhing turtle in my left hand while shifting and steering with the other, I careened down the winding river road.

After narrowly missing several trees, I parked the car near the river, ran down somebody's stairs and dropped the turtle under a boat dock. The water there was very still, and so, at last, was the Stinkpot. Despite all the handling, however, my Stinkpot never stank—not in my hand, not in my home, not in my car. But I no longer believe the stereotype of the slow-moving turtle. If the tortoise had been a turtle in the story of the big race, my bets would not have been on the hare.

❧ ❧ ❧

TURTLE TALK

Descriptions of turtles in guidebooks throw some unusual terms around as if anybody past third grade would know them. I didn't, and I couldn't make heads or tails of turtles until I did:

Carapace:	The top shell
Plastron:	The lower shell
Bridge:	The bony extension on each side that connects the plastron to the carapace
Keel:	The ridge down the carapace (or the plastron)
Scute:	The plates that make up the carapace

OCTOBER 2

Ladybug Beetles

The ladybugs are landing! They are everywhere: on feathers, driftwood, sand, on and under rocks. Huge numbers of ladybugs gather on the beach every October, but no one around here seems to know why. Stumped, I called Earl Wolf.

The best explanation Earl had ever heard for the beetles' fall arrival on the beach was migration: "These are some of the few beetles that hibernate as adults, but we know they don't winter on the beach," he said. "They just don't like to fly over water. Nearly all beetles can fly, but their wings are so small, they can't manage much of a wind. When they get to the lake, where it's usually windy, they stop and gather until conditions are quiet enough to move again." When I expressed surprise that beetles migrate, Earl said, audibly amused, "Migration doesn't have to mean Argentina! Migration can mean Indiana, or even a few miles south."

Ladybugs like to cuddle up for winter: In California, millions of them have been found en masse. Ladybugs diet on aphids mostly, which make them beloved by farmers and gardeners everywhere. They were named for "Our Lady" during the Middle Ages when they "miraculously" rid grapevines of destructive insects.

☙ ☙ ☙

HOW MANY SPOTS HAS A LADYBUG GOT?

Sometimes none, sometimes a lot. Some common North American ladybugs are the Two-Spotted Ladybug, the Nine-Spotted Ladybug, and the Spotless "Nine-Spotted" Ladybug, a subspecies of the Nine-Spotted variety which, surprisingly, has no spots at all. Ladybugs come in different colors, too: red, orange or yellow with black spots, for example, or black with red or yellow spots. Generally, the number of spots helps identify a species, but there are many species—over four thousand worldwide.

OCTOBER 4

Witch Hazel

It was love at first sight when I met Witch Hazel, even before I knew its eccentricities. A couple of weeks ago, I was walking along an unexplored piece of road looking at leaf edges when a simple, lopsided, egg-shaped leaf caught my eye. So far, I'd noticed toothy, smooth and serrated leaf edges, but these edges were wavy, almost scalloped. I suppose if I hadn't been looking for odd edges, I'd have walked right by. At home, I identified it immediately; a Witch Hazel leaf is unlike any other.

Today, at the top of a high dune, I found the shrub again, this time in bloom, its thin, dark branches flowering in shocks of spidery, yellow petals, looking very like a sparse forsythia. Witch Hazel is often the last bright bloom of the year, because the flowers can curl back into their buds when the weather is cold, to burst out again in warmth. I think it odd that the first and last blooming shrubs should look so much alike.

Witch Hazel was named long ago for its unconventionality. Forked Witch Hazel twigs have long been used as divining rods by "well-witchers." Furthermore, not only does Witch Hazel bloom unexpectedly, but its woody pods spit their contents with explosive strength. My favorite turn-of-the-century author, Neltje Blanchan, suggests in *Nature's Garden* that Witch Hazel pods, brought indoors to ripen, appeared to use living room contents for "small artillery practice," pitching seeds as far as twenty feet.

NATIONAL BIG TREE REGISTER

Michigan claims the Witch Hazel Big Tree—the biggest of its species in the nation—located less than an hour from me, in the Muskegon State Park. Michigan has ninety Big Trees listed in the *National Register of Big Trees* (see Richard Pardo, Bibliography), which reports the location of the largest specimen of each species of tree in this country. Florida is the only state with more.

OCTOBER 6

Blue Racer

I was getting impatient with the paucity of wild-life I was seeing at the lagoon, which I was showing off to a friend—in contrast to the Mute Swan excitement of last spring, the most we'd seen were tiny olive-colored frogs plopping like raindrops into the shallows—when I nearly stepped on a silky, blue-gray snake. Zip! It vanished with the merest parting of beach grass. I recovered and charged after it, too late. "Damn, that thing was *fast!*" I said to my companion, who was still somewhat in shock. We tried to remember what we had glimpsed: several feet of blue-gray lightning. Not far down the path, we heard another hiss of grass and saw the bank tremble slightly as another snake slipped away. Finally, a third snake offered us an extra split second for details: round as rolled clay, smooth and evenly colored, four or five feet long and oh so fast.

Later we easily identified the snakes as Blue Racers. No other snake looks quite like a Blue Racer. Even though the color varies by area—gray, bluish or greenish—most people will not be confused. Three to six feet long, round-bodied, slender and swift, with a tough diet, including insects, birds, frogs, toads and other snakes, the Blue Racer is distinct.

❧ ❧ ❧

RACERS: NOT MUCH COMPETITION

Other racers, similar in build and speed but different in color (as indicated by their names), are common, but the ranges of the snakes seem not to overlap too much.

Midwest:	Blue Racer
Central and Northeast:	Northern Black Racer
Southcentral and Southeast:	Southern Black Racer
Central:	Eastern Yellow-bellied Racer
Rockies and West:	Western Yellow-bellied Racer

OCTOBER 7

Monarch Butterfly

Last weekend I was bird-watching in a nearby apple orchard where milkweed grows in the long grass, when I was struck by a new proliferation of Monarchs. Newly hatched, their brilliant wings were unfaded, fresh and strong. They flew, in that buoyant way of butterflies, between the rows of crooked trees hung with scattered last apples, the reddest apples I have ever seen. Orange butterflies, crimson apples, warm blue sky, long, lush grass crushed by the bodies of last night's sleeping deer—it was a delicious autumn afternoon.

So new and innocent were the Monarchs that I could approach them as they clung to bright balls of purple clover, slowly moving their damp wings. Although the Monarchs were easily caught between my thumb and middle finger, their slender legs were harder to remove from the blossom than a burr from a sock. Squatting only a foot away, I studied the exquisite veining, that incredible paradox of delicacy and strength that soon would carry them thousands of miles.

The Monarchs are everywhere now. At lunch recently in a neighbor's patio, one of them circled us for about ten minutes, and then lit in my hair—a strange tickle of tiny legs—before sailing over the roof, toward the lake. Early this morning, I saw one flying all alone, sun flashing off orange wings over flat, green-violet water. They were the only wings visible in the vast, clean sky. The orange-and-black Monarch is easily identified, unless you are fooled by its mimic.

Much has been made of mighty Monarch migrations, sometimes over thousands of miles. No single butterfly, however, makes the flight both ways; they breed once in the North, and again when they begin their northern migration in the spring. Only the offspring make it all the way back.

☙ ☙ ☙

VICEROY: THE MONARCH TWIN

The best way to tell the slightly smaller Viceroy from the Monarch is by the black cross-stripe on the Viceroy's hind wing. This requires a fairly close inspection. Mimicking the Monarch coloration is how the Viceroy preserves itself: Because they feed on Milkweed (see page 173), Monarchs taste bad, so are not so much preyed upon.

OCTOBER 10

Honey Bees

This afternoon my neighbor Al invited me to help him "work" his bees. We followed a path through brambles and sumacs turned orange and rose, through yellow-leaved shrubs with shiny, oval, red berries, into a small clearing where three hives, each a stack of three boxes about the size of desk drawers, called *supers*, stood like misplaced file cabinets. Inch-long, amber-colored bees drifted through the air.

Al stuffed some dead leaves in a little can with a funnel cap, which was attached to a small bellows. He lit the leaves, worked the bellows and out came smoke. "You'll be the smoker," said Al, handing the device to me. At his instruction, I pumped some smoke through the bottom entrance of each hive. "When the bees smell smoke, they worry that the hive is on fire," explained Al. "Their first impulse is to move the hive—including as much honey as they can. So now they're very busy gorging themselves. Bees with full abdomens are less aggressive."

While the bees were banqueting, we armored ourselves. Al fit a netted wicker safari hat on my head. I put on heavy gloves that came over my elbows and tied my boot laces around the bottoms of my pant legs. Al himself, shirtless until now, no doubt to prove his claim that bees are docile, donned a thin cotton shirt, rubber bands on his pant legs, a smaller netted hat and a pair of short rubber gloves.

I tensed as we approached the first hive. Al began a soothing, engaging patter that he continued the entire time we were there, about bees and their habits and ways. He lifted off the top of the first hive and instructed me to puff smoke over the top. "Gently!" he corrected, as I pumped the bellows. "We don't want a gale!"

I looked down on a moving mass of bees, thousands of them, crawling between the ten frames that fit neatly inside the super. Each frame held a thin wax sheet imprinted with a honeycomb pattern, on which the bees built wax cells for honey or eggs. Al carefully slid out one of these frames. ("More smoke!") Wax cells had been built about an inch deep on each side. "If I were harvesting this honey," said Al, "I would slice off the caps with a sharp knife and spin the frame in a centrifuge." He had recently harvested about eighty pounds per hive and now was checking to see if there was enough new honey to get the bees through the winter. There was. Many cells were plugged, meaning they were filled with honey. As Al slipped the frame back, I puffed more smoke on mounding bees to thin them, lest they be crushed, and he checked all the remaining frames, each crawling with hundreds of amber-brown bees.

To get at the middle super, Al lifted the top one off—*"Smoke!"*— and set it in the grass, and we went through the process again of checking each frame for honey and healthy bees. The bees were becoming more active. "More bees are landing on me than on you!" I complained. "That's because they know me," said Al. "They're curious about you."

By the bottom super—the hive was now completely unstacked—bees were becoming agitated. They were everywhere. I kept pumping smoke on bee clumps and Al kept talking. "That bigger bee is a drone," he said. "The drones are the males, whose only purpose is to impregnate the queen. When winter comes, the worker bees—all sterile females—will throw out all the drones. And see there? A worker is carrying off a dead bee! Bees are incredibly clean. They are impeccable. They leave the hive in order to get rid of bodily waste. If you come here in winter, on a warm day, you'll see all sorts of yellow spots on the snow."

Suddenly we noticed activity at the bottom entrance. Al explained that there were guards there. "Bees never know when some foreign bee might sneak in for some easy pickings. Each bee has the smell of its own queen,

> ## WHERE TO SEE A HONEY BEE
>
> Look in a nice big patch of flowers for a less-than-an-inch-long, amber-brown bee with black and soft amber-brown rings on its abdomen. Although there are other amber-brown bees, they aren't nearly as common as Honey Bees. That "bee" buzzing around your head probably isn't one (see page 273): Honey Bees are far too busy to bother with you unless you are seriously threatening the hive. It takes fifty thousand bee miles—nearly twice the circumference of the earth—to make one pound of honey.

however, so foreigners are soon thrown out." We saw worker bees returning there with a golden ball of pollen on each of their hind-leg knees, collected in a pocket called a "pollen basket." Pollen is eaten, too, especially by developing bees. "The strength of the hive is completely dependent on the queen," Al said. "If you get a good queen, you have a good hive. A good queen can go on for several years before another queen replaces her."

Al restacked the hive and topped it with an extra-roomy cover for winter. He lined up the supers exactly, to keep out the mice. Then we repeated the whole unstacking and checking procedure for the two remaining hives. When we had finally stacked the last hive—"Extra smoke!"—Al pushed my ear inches from the thick, crawling carpet of bees—"Fifty thousand of them!" Deep rumbling vibrated in my stomach. Then we covered the last hive, got our equipment together and walked back through the sumacs. Neither of us had been stung once.

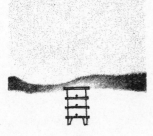

OCTOBER 12

The Virgin Queen

"Did I tell you about the sex life of the virgin queen Honey Bee?" asked Al on the way back from the hives. No, I didn't believe he had. I certainly would have remembered this one.

When the old queen Honey Bee stops laying eggs properly—leaves empty cells or makes too many drones—the worker bees kill her. In the spring, the first virgin queen to hatch begins her reign by tearing open the cells of the other potential queens—queens develop in special, extra-large cells—and killing her competition. The virgin queen then waits a few days for her wings to develop. Meanwhile, drones from various hives begin hanging out together on treetops, waiting to spot a queen and mate with her. Drones have extra-big eyes, all the better to see a queen with, for when she finally takes to the air, she flies very high. This makes it most possible for her to mate with drones from many different hives, thus avoiding incest and assuring strong offspring.

After a few days of orientation flights and a three- to sixteen-day rest period, the virgin queen spirals high in the sky, where she is found by a drone and mates with him. For the drone it's a love death. Once the queen has his sperm, she shakes him loose and the drone is literally torn apart, his abdomen remaining attached to the queen, the rest of him falling a good hundred feet to the ground. The queen flies back to the hive, trailing proof of her accomplishment. She is welcomed by the sterile female worker bees, who become very excited, showing their pleasure by licking her, removing the evidence, and preparing her for an encore flight about an hour later.

The queen makes five or six of these flights, and then she retires permanently to the hive, heavy with sperm from five or six drones. She stores all the sperm in one part of her body, and her eggs in another. The rest of her life she puts these together, inserting fertilized eggs into worker-made cells, making hundreds of thousands of bees every year, for as long as five years. She never mates again.

OCTOBER 14

American Bumble Bee

Today I walked down a path padded with butter-yellow Sassafras leaves—softly rounded goose feet, mittens and eggs—past red-and-orange sumac, little brown bird's nests of dried Queen Anne's Lace, an occasional bright blue Chicory blossom that had escaped the mower, and along puffy plumes of shoulder-high Canadian Goldenrod gone to seed. After a bit, I came to a table-sized patch of unimpressive, small, white asters blooming on knee-high plants at the edge of a pasture. Although it's not very showy, beekeepers love the Small Aster (also called a Snow Aster), for it blooms longer than most other flowers do, providing bees with a late source of nectar for winter.

This was a busy patch, so I sat down in the middle of it. The flowers now at eye-level, insects lit inches away from my face, but none of them paid any attention to me; not one of them flew around my head, or lit on me, or seemed to care one buzz about my presence there. I observed several Paper Wasps: needle-waisted, needle-tailed, mahogany and black, with yellow rings around the abdomen. Their wings were dark mahogany, too. Some had dark faces, some light. (I read later that the light-faced male Paper Wasps have no stingers.) I watched bee tongues probe deeply into the flowers, and pollen collect on their knees.

Two very different kinds of bees were amiably sharing the loot. The Honey Bees were the familiar ones, slimmer, amber colored, almost brown. (I noticed that the Honey Bees and the bumble bees flew off in opposite directions.) The larger ones were American Bumble Bees: fuzzy, tawny yellow, with big abdomens that began yellow and ended in black tail tips. The bumble bees were so furry I wanted to pet them, put them on a leash and take them home. I moved softly now and then, pulling a stem toward me so I could better observe its visitor. There were two sizes of bumble bees: large queens and smaller workers. I wondered what queen bumble bees were doing out in the fall.

HOW MANY TIMES CAN YOU BE STUNG?

It's rare to be stung, but if you are allergic to the venom or have never been stung, be cautious: reactions can be serious. Usually only female bees, wasps and hornets sting. Honey Bee workers sting once, then die. Bumble bee workers live on to sting again and again, as can Yellow Jackets and Bald Faced Hornets. Paper Wasps are repeaters, but they are less aggressive than the hornets.

🐝 🐝 🐝

WATCH FOR QUEEN BUMBLE BEES IN SPRING AND FALL

There are several kinds of very similar bumble bees in North America; one of the biggest, the American Bumble Bee (also called the Common Bumble Bee) showing up in most of the United States and southern Canada. Unlike Honey Bee queens, bumble bee queens spend a good deal of time outside the hive and mate in the fall. These newly fertilized queens are the only members of the hive to survive the winter. In spring, they can be easily recognized, the largest bees around, buzzing loudly from flower to flower, busy beginning a new bumble bee colony in some underground den. Although bumble bees make honey, they make far less than Honey Bees, which must feed a whole colony through winter.

OCTOBER 16

White-winged Scoter

Early this morning, I was looking for migrating ducks along the shore when I came upon a huge, bat-winged, rainbow kite hanging motionless over the surf, no person in sight. Using my binoculars, I followed about a hundred feet of string to a large tree, halfway up the dune, that appeared to be flying the kite from its tip, where the end of the string was tangled. Hoping to recover it, I waited for the kite to come down, but it continued hovering brilliantly, eerily, over languid waves.

After ten minutes, I continued on, and about a quarter of a mile down I found a tight flotilla of dark brown, gull-sized ducks floating about fifty feet offshore. They did not spread out and play like mergansers and Buffleheads; instead, the twenty or so birds swam so close together, they could have fit in a bathtub. The brown females and immature birds had a large white spot on either side of each eye; the darker males had a funny black bump over each of their squat, orange bills. Every few minutes, the entire group would vanish in a simultaneous dive, like an Esther Williams troupe, to feed, I have read, on mollusks up to twenty-five feet below the surface. Emerging together, they'd close ranks, solid as a boat.

They resembled the Surf Scoters in my guide, but I took them to be White-winged Scoters, which, according to the range maps, is the only species of scoter to frequent the Great Lakes. Tight floating formations are also typical of the White-winged variety. I was thrilled to see a scoter, never having heard of one before.

After a quarter hour of scoter water ballet, I headed back down the beach and found the kite floating about thirty feet out, the string now just within reach. I pulled the kite through the water, like a big fish, leaning my whole weight back on the string. When I finally severed the unbreakable string with a lit match, the beautiful kite was mine. I carried it home high over my head, its lift so strong I almost flew.

᠍ ᠍ ᠍

THE DOUBLE-DOTTED DUCK

A brown duck with two large, white spots on each side of its head is likely a female or immature White-winged or Surf Scoter. Seen on the Great Lakes, it's probably the first. You can tell the White-winged Scoter from the others by its white wing patches, especially in flight, or from the black bump that tops the male's orange bill. Both these and a third scoter, the Black Scoter, summer in Canada and Alaska and winter along the East and West Coasts of North America.

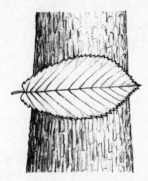

OCTOBER 17

Eastern Hop Hornbeam

On my way home from the beach yesterday, between the scoters and the kite rescue, I discovered a slim tree that was surrounded by birchlike leaves, but had bark—vertically shaggy and shingle-like—that was quite unlike a Paper Birch's peeling, thin skin. I took some leaves home and discovered that there really aren't that many kinds of trees with doubly serrated, oval, pointed leaves, and, as luck would have it, the bark on each of those that do is quite distinctive.

ᘓ ᘓ ᘓ

DISTINGUISHING TREES WITH BIRCHLIKE LEAVES

In the West, the only trees with simple, leaf-shaped, doubly serrated leaves besides birches are alders, and most tree-sized alders are along the West Coast or in the Northwest. In the eastern half of the United States, there are four groups of trees with this type of leaf: birches, elms, hornbeam and Hop Hornbeam (Eastern alders are usually shrubs or small trees). These are easily distinguished by their unusual bark:

TREE	BARK AND TRUNK
Elms	Furrowed; trunks can grow large
Birches	Thin, papery, sometimes peeling; trunk fairly smooth
American Hornbeam	Thin, gray; narrow trunk looks ropy
Eastern Hop Hornbeam	Shaggy, like thin, vertical shingles; trunk narrow

OCTOBER 18

Shaggy Mane

It's been raining so much lately that mushrooms are shooting up along my daily exercise route overnight: white soldiers high as my hand, shaggy caps not yet opened, still locked near the bottom of the stem. These have turned out to be Shaggy Manes, a fairly easy mushroom to identify. I was surprised to learn that several of my friends collect Shaggy Manes, considering them a fungal delight second only to springtime morels, a delicacy I've never been able to find.

A mushroom similar to the Shaggy Mane is popping up in the lawns as well. It opens more quickly, like a Japanese umbrella, exposing inky black gills that almost immediately begin dissolving into a black puddle, oozing black "blood." Shaggy Manes and Inky Caps, although quite different from other mushrooms, do resemble each other.

I haven't eaten my Shaggy Manes. Mushrooms really can be very hard to identify, remaining only briefly in their prime. My beginner's luck with these two didn't hold; mushrooms I've found since either looked like many of the guidebook pictures or like none of them at all.

🐌 🐌 🐌

SHAGGY MANE OR INKY CAP?

Both Shaggy Manes and Inky Caps peel, shinglelike, around the cap and have gills that turn black with age. Shaggy Manes are edible when the gills are young and white; Inky Caps are edible but are reputed to be much less tasty and can cause problems if alcohol

---

TO EAT OR NOT TO EAT

Even at peak freshness, edible and poisonous mushrooms can be easily confused, with the deaths of mushroom "experts" occurring every year. Mushroom guides strongly recommend that identifying wild mushrooms for consumption should be left to the experts, and that no one should ever eat a wild mushroom raw. Neither the author nor the publisher will take responsibility for the consequences of wild-mushroom consumption.

is consumed before, with or after eating them. To tell the difference, look for the loose ring sliding up and down along a fresh Shaggy Mane stem. A young Shaggy Mane cap is oval and almost, but not quite, grasps the stem.

MUSHROOM OR TOADSTOOL?

What's the difference between a mushroom and a toadstool? There isn't any. I always thought a mushroom had gills, while a toadstool was a fungus without gills. I was wrong. Mushrooms fall into three groups: 1) nongilled mushrooms, like morels; 2) gilled mushrooms, like the Shaggy Mane and the Inky Cap; and 3) puffballs and related mushrooms. *Toadstool* is a folk term, often referring to any mushroom that is poisonous.

OCTOBER 19

Honey Mushroom

My mushroom books suggested consulting a mushroom expert, so I arranged a mushroom-hunting party and invited some friends and a couple of amateur mycologists. It turned out to be the stormiest fall day yet: wind screaming, trees frantic, ten-foot waves pounding the beach, rain rattling on the windows, small hail balls binging off the panes, temperature dropping into the forties. I feared my grand plan—that my guests and I would pad through the dappled autumn woods, poking at the forest floor—was doomed, but I hadn't counted on the passion of mycologists. "Mushrooms thrive in wet weather, so what are we waiting for?" All but three of us ended up braving wind, hail and rain in search of the wild mushroom, each of us toting a small bowl and paring knife.

First, we found Inky Caps, which Russ, the most knowledgeable among us, referred to as *Coprinus atramentarius*. Russ rolled this off with the ease of calling his daughter home for supper. It's safer to use the proper Latin names, he explained. You don't want to make a mistake. Mostly, though, we saw Honey Mushrooms (*Armillaria mellea*), squat, amber-capped fungi with brownish stems, massing tightly around the bases of large trees, as if mobbing a vampire's castle. If we hadn't known where to look, we'd have missed them, they were camouflaged so well in unraked oak leaves.

Mushrooms are fungi: leafless, flowerless, rootless, seedless plants which, lacking chlorophyll, feed on nutrients in wood or soil. Some mushrooms have a *symbiotic*, or mutually beneficial, relationship with the tree roots they grow on: The mushrooms provide the tree with minerals and help in times of drought, while the tree provides the mushroom with nutrients. Other fungi break down dead material. Parasites, they kill the material they feed on. Honey Mushrooms take their sustenance from tree roots and appear to be of the parasitical variety, sometimes harming the trees. Mushrooms are often picky about which kind of tree they prefer. Honey Mushrooms like hardwoods, especially beech and oak.

Russ dug some out with a small knife. "You don't destroy the plant by taking the mushroom," Russ said. What we think of as a mushroom is only a small part—the "fruiting body"—of the mushroom. Mushrooms are made of *mycelia*, or filaments. Most of these live underground, unseen, but under certain conditions, they "fruit" above ground, releasing *spores*, from which other mushrooms can grow.

ᴈ ᴈ ᴈ

THE SECRET SEX LIFE OF MUSHROOMS

For mushrooms to reproduce, spores from fruiting mushrooms must land somewhere, develop and then search through the soil for another spore. After an underground, unseen union of two primary mycelia, each having developed from the same spore or, more commonly, from different spores, a fruit can be made. Mushrooms, the "fruiting bodies," are "secondary mycelia."

OCTOBER 20
Puffball

I went mushroom hunting again, hoping that yesterday's rain would provide me with some samples. The woods seemed to smell of mushrooms, rich and damp. A deer crashed away through the sunny Sassafras; a Hairy Woodpecker popped in and out of a hole high up on a snag. I recognized Sugar Maples, Sassafras, Eastern Hemlocks, Basswoods, White Ashes, Staghorn Sumacs. I don't know if a particularly astonishing number of mushrooms appeared today (I have walked this forest many times before; how could I have missed them?), or if I am simply learning where to look, but mushrooms seemed to be everywhere. Small, delicate, fawn-colored mushrooms grew up the trunks of living trees, anchored deeply in the crevices of bark. I checked around dead trees, where I found Honey and other mushrooms, including exquisite bright orange ones, with spaghetti stems and caps like pennies. Dead logs offered white shelf mushrooms and delicate, striped, brown fantails called Turkeytails. I found little puffballs sprouting like Ping-Pong balls on rotting two-by-fours and big, whitish, pancake-topped gill mushrooms growing near a summer cottage.

I was really looking for puffballs, but the only one I found was the size of a baseball, not too fresh, with dirty-looking skin beginning to crackle. Split, it looked like chocolate sponge cake inside, and when I squeezed it, brown spores puffed out like smoke. I've heard that puffballs are delicious, so I assumed that a puffball was a puffball, but apparently not. There are several pages of puffballs in the guides and not all are edible. Many puffballs are edible when young and white—some grow big as basketballs—but the books advise cutting them in half to check for any developing gills or caps that could make them a poisonous species. Young puffballs can be confused with developing and deadly amanitas (see page 271).

Of the fifteen kinds of mushrooms I brought home, I could only identify a few, although I spent hours paging through four different guides. My puffball was one of them; I think it was an aging *Bovista pila*.

✿ ✿ ✿

HOW TO GATHER MUSHROOMS

Mushroom hunting is big in Europe. There are probably more mushroom hunters in Czechoslovakia than there are birders in Britain! Here it's catching on, but slowly. Some advice for beginners:

1. Pick fresh ones; only a fresh one will give all the proper clues for identification.
2. Be sure to dig out the whole thing, including the base.
3. Use waxed paper or aluminum foil (plastic bags cause mushrooms to decay).
4. Keep the mushrooms separated from one another.
5. Use a shallow basket; mushrooms are fragile.
6. Don't eat any wild mushroom raw.
7. *Don't eat any wild mushroom that has not been positively identified as edible by a mushroom expert.*

OCTOBER 22

Death Cap

When a friend and I took a fall walk along the Swan River yesterday, my eyes searched among the fallen leaves; I was still obsessed with mushrooms. The most distinctive we saw was a large, luminously lemon-colored mushroom with pristine white stem and gills and a waxy, shiny cap.

It was a perfect, fresh, delicious-smelling miracle, practically newborn; we could even see the egg-shaped structure, large and bulbous, from which it had grown. I picked it and brought it home.

This morning I identified it as a Death Cap (*Amaninta phalloides*) possibly the most poisonous mushroom in the world; one bite can kill. It also could have been the False Death Cap (*Amanita critrina*) less lethal, but still poisonous. The shiny yellow cap is more typical of the False Death Cap, but there are also cottony patches on the guide's False Death Cap, and there weren't any on this one. There always seem to be these problems with mushrooms, which require more expertise to identify than I have. I think I have something identified, and then some characteristic is mentioned in passing that just doesn't apply. The mushroom could have been a Death's Cap, which has been found, sometimes by the bushelful, in the eastern United States. Its range is reportedly increasing, and any mycologist can expect to find it in his or her range.

I worried for days that, possibly having handled a deadly poisonous mushroom, I inadvertently might have ingested some small part of it, a bite of which could slowly destroy my liver. I find it odd that I haven't found a warning about handling lethal mushrooms anywhere in my extremely cautious mushroom guides, and my amateur mycologist friends assure me, with contained amusement, "You have to eat some to be poisoned. You can't absorb the poison through your hands." I'm not convinced.

⁖ ⁖ ⁖

WATCH OUT FOR THE AMANITA "EGG"!

A mushroom that grows out of a bulbous egglike structure, called a *volva*, is likely a species of *amanita*, not a nice family to have to dinner. None of them are recommended for eating; most are poisonous, some lethal.

OCTOBER 23

Yellow Jacket

Everyone was complaining today about the "bees" at the Goose Festival, a street fair in a neighboring town that celebrates the arrival of thousands of Canada Geese, which winter in a nearby preserve. The fair-crashing insects buzzed around the trash cans, flew dizzy circles above food spilled in the street, tried to crawl through the small holes in sugar dispensers at street food concessions and drove me cross-eyed as I hurried to eat my pizza slice before I got stung. The little buzzers were about the same size and shape as Honey Bees, but I got a very close look at their slick, meticulously patterned black-and-yellow bodies as they attacked my pepperoni, and I realized that they weren't bees at all. They were Yellow Jackets, which are hornets, not bees. Yellow Jackets adore "people" food. Bees eat only nectar and pollen.

Yellow Jackets are especially aggressive in the fall, becoming common, obnoxious picnic guests. Begun in spring by a single fertilized queen (unlike Honey Bees, which start out with a full complement of workers and drones), this colony probably has now multiplied to more than twenty thousand. With winter coming, Yellow Jacket workers are no longer busy raising young and are free to get into mischief. Bees are such sweetie pies that it's a shame to confuse them with Yellow Jackets.

ᴣᴀ ᴣᴀ ᴣᴀ

HOW TO TELL A HONEY BEE FROM A YELLOW JACKET

Any "bee" buzzing around your head is probably a Yellow Jacket. It's easy to confuse Honey Bees with Yellow Jackets, for they are about the same size (three- to five-eighths of an inch) and shape. Honey Bees are far gentler than Yellow Jackets, though. Here's what to look for:

HONEY BEE	YELLOW JACKET
Fuzzy body	Nearly hairless
Amber-colored with black rings	Black with bright yellow patterns
Direct flight style	Frequently zigzags

Daddy-Long-Legs

I was about to step into the shower, when I noticed I had some slim, leggy company. A daddy-long-legs—a really big one—was trying to escape the tub. The eight long, threadlike legs couldn't get a grip on the slick enamel. I watched for a while, then gently picked it up and placed it on my sleeve. It did not drop on a silken thread. Instead, its round, three-eighths-inch-wide body hunkered down low, like the center of a flower, while each delicately segmented leg jackknifed around it. Then it ferociously waved two of its legs at me.

I wasn't frightened. So many large, scary eight-legged creatures crawled through my Pakistani childhood, including scorpions and tarantulas, that a daddy-long-legs has always seemed appealingly harmless. It doesn't bite, rarely runs and tolerates long observation. At worst, it lets loose from stink glands behind the first pair of legs. A daddy-long-legs's fragility is also its defense: It thinks little of shedding a leg for its life, even though a lost limb cannot be regenerated.

I released the daddy-long-legs in another part of the house; maybe it would catch some of those annoying fruit flies that recently invaded my kitchen. Because I like daddy-long-legs, I had always prided myself on not being afraid of spiders. But I was wrong.

🐜 🐜 🐜

WHEN IS A "SPIDER" NOT A SPIDER?

When it's a daddy-long-legs. There are ten orders of the *arachnids* (eight-legged "bugs"), among them spiders, scorpions, mites and daddy-long-legs. Although a daddy-long-legs is an arachnid, it is *not* a spider, but a *phalangid*. Body build is one difference. Several sources indicated that daddy-long-legs have only one obvious body part. My find, however, seemed to have a tiny head as well. Here are the differences:

An insect body has three parts: head, thorax and abdomen.
A spider body has two obvious parts: a head/thorax combination called a
cephalothorax and an abdomen.

A daddy-long-legs body has one obvious part: It does have a cephalothorax and abdomen, but no "waistline" separates them.

LOTS OF DADDY-LONG-LEGS

There are over three thousand species of daddy-long-legs in the world, and over two hundred of them in the United States. Of these, four are common.

OCTOBER 27

Garden Orb Spider

I just read that in late fall, especially following a cold snap, spiders creep indoors. A quick inspection of my window frames confirmed this and explained the sudden silk that has begun to fuzz up the view. I don't really mind spiders around the house. I found two kinds of spiders parked in the window frames. The first played dead, legs suddenly shrinking up close to its body, crabbing out, four on a side. It was the only spider I've ever found an exact match for in my spider book—a dead ringer for the photo of a Garden Orb Spider. Large body (half an inch), with a quarter-sized leg spread, it was probably female, as the males are so tiny that they can climb over the female's body. The legs were striped, black and pale yellow. The pale yellow body was marked delicately with black, especially on its tear-shaped abdomen, which came to a point at the end. The Garden Orb Spider is famous for its delicate, circular web.

A second spider dangled before the glass on a yard-long length of silk. It was gray and tiny, less than a quarter inch, legs and all, with the ballooning abdomen typical of most house spiders. Both my spiders were harmless to humans, as are most of their kin.

🐜 🐜 🐜

ONLY TWO POISONOUS SPIDERS HERE

Of the approximately two thousand spider species in the United States, only the Black Widow Spider and Violin Spiders are poisonous. (Despite its monstrous appearance, the imported pet-store Tarantula can't kill much more than a beetle!) The female Black Widow—only the female bites—is quite easily recognized by her fairly large (half-inch), shiny, black body; since we have few shiny black spiders. (If you'd care to turn her over, you'll find a scarlet warning pattern.) Her venom attacks the nervous system and can cripple the whole body, with a mortality rate of about 4 percent. The Black Widow, known to love outhouses, can be found all over the United States.

The several species of Violin Spiders are more difficult to spot: A medium-sized (quarter-to-three-eighths-inch body) brown spider, its most unique feature is the arrangement of its six eyes (most spiders have eight): The three pairs appear as three black dots on the top of its head. Violin Spider poison works locally, and although not lethal, it can attack the flesh around the wound and cause an ulcer that can take months to heal. Violin Spiders prefer the Midwest and western part of the United States.

OCTOBER 29

Turkey Vulture

I saw five or six Turkey Vultures circling wetlands near here, harassing seven swans, who paid them no attention. Buzzards, I thought, with distaste. When I looked it up, I read that the Turkey Vulture is not a buzzard. This was startling information. I have always called it a buzzard, but *buzzard* comes from *buteo*, Latin for a kind of hawk or falcon. Early settlers mistakenly called our vultures "buzzards" and the name stuck.

Known for their love of carrion, vultures are common worldwide. I have vivid childhood memories of watching vultures lined up on tree branches outside our Pakistani home, their bald, turtly heads hunkered forward, contemplating, I was certain, their next horrible meal. These were the vultures of nightmares, far uglier than our turkey version, which ranges just about everywhere in the United States. And what's so bad about eating dead meat? Bald Eagles and humans do it, too!

It's fairly easy to recognize a Turkey Vulture with the aid of binoculars:

Its large size, bald head and black-and-white underwings make it distinct. Still, I have wondered for years how to know one when it's soaring.

꙰ ꙰ ꙰

THE TELL-TALE DIHEDRAL

A large, dark bird soaring with wings cocked in a *dihedral*—a slight V—is probably a Turkey Vulture. Bald Eagles and most hawks—even the Black Vulture of the Southwest—soar flat, or almost flat. In the East, the only other dihedral-soaring raptors are the smaller, slimmer, lighter Northern Harrier and the fairly rare Golden Eagle. Most raptors that soar in a dihedral are much smaller than the Turkey Vulture.

Eastern United States:	Northern Harrier, Golden Eagle
Western United States:	Northern Harrier, Swainson's Hawk, Ferruginous Hawk, Golden Eagle
California:	Northern Harrier, Black-shouldered Hawk
Gulf Coast:	Northern Harrier, Black-shouldered Hawk
Deep Southwest:	Northern Harrier, Zone-tailed Hawk
South Texas:	Northern Harrier, White-tailed Hawk

OCTOBER 31

Northern Catalpa

This morning I found that a Halloween trick had been played on me by the huge Northern Catalpa that shades the driveway: My van was nearly buried under its huge, heart-shaped leaves. The Northern Catalpa is so sensitive to cold, that when hit by a hard freeze, it dropped nearly every last one of its still green leaves in one dramatic dump. Only yesterday, it was lush with a copious canopy; now it stands naked. The Northern Catalpa has other dramatic capabilities. Not only does it bloom in early summer with huge, white, purple and yellow orchidlike flowers, but in fall, before its leaves fall, it drops an impressive number of foot-long seed pods.

It was while studying sample leaves of this tree that I discovered the delicious feel of some leaves; the huge, smooth-edged, heart-shaped Northern Catalpa leaf feels like baby's skin on top, downy soft beneath. The tooth-edged, slightly lopsided Basswood leaf, another heart-shaped leaf, has large veins with thin, delicate green skin stretched between them.

ᵂ ᵂ ᵂ

COMMON TREES WITH LARGE, HEART-SHAPED LEAVES

Large, heart-shaped leaves are easy to spot and only a few common trees have them:

Northern Catalpa has the largest leaves—up to a foot long and seven to eight inches wide—smooth-edged and opposite.

Redbud leaves are smooth-edged, too, but only half the size of Northern Catalpa, and alternate.

Basswood (American Linden) is distinguished from Northern Catalpa and Redbud by its saw-toothed leaf edges.

November

Ball Gall

I've seen almost perfectly round growths right in the middle of some goldenrod stems which looked like plant tumors. They turned out to be hybernation dens. Goldenrod, among many other plants, are frequent hosts to insect larvae. A small fly, for example, lays its eggs on a goldenrod stem in late spring, and the hatching larva digs into the stem and hollows out a nice little "womb" for itself. The plant slowly balloons around it, growing a round ball gall. The fly larve takes as long to develop as a baby, staying in the gall all winter (unless it's invaded and eaten by some other insect) and coming out the following spring as an adult. Folks who ice fish sometimes cut open a ball gall and use the grub for bait.

People who fish aren't the only ones aware of galls. A gall is often opened by a bird—some woodpeckers are experts!—for food, or by another insect, which may find it a convenient winter home.

❧ ❧ ❧

GALLS FOUND ON DRIED FALL AND WINTER PLANTS

There are many kinds of galls to be found in fall and winter, each made by a different sort of insect, which is often quite particular about the type of plant it invades. Here are some other common fall galls:

Elliptical Galls: Longer and more slender than ball galls, also found on goldenrods, and inhabited by the Ichneuman Wasp.

Leafy Bunch Gall: Misshapen leaves at the end of a branch, usually on Canada Goldenrod, made by a midge and usually empty by winter; looks like a flower.

Pine Cone Willow Gall: Made by a fly on willow branches, looks like a tightly closed pine cone; green in spring, then yellow to brown and finally gray in winter.

NOVEMBER 4

Norway Maple

When I drove into Grand Rapids today, our nearest big city, I was astonished at the brilliance of the Norway Maples. Several days of rain had soaked their tree trunks black, and, crowned with gold against a new blue sky, they provided a colorful contrast to a recent spate of damp November days. I had seen a few of these bright yellow trees, brilliant as goldenrod, scattered in our woods, but in the city they made golden tunnels of the residential streets, their color splashing onto lawns, glowing gloriously in parks. Norway Maples are not native trees, but were imported from Norway via England, so they are more widespread in cities than in the woods, where they sometimes stray from residential properties. A Norway Maple leaf, wider than tall, is most easily distinguished by the milky substance that oozes from a broken leaf stem. They don't ooze much in fall, but the late leaf color is a glorious giveaway.

ɘ ɘ ɘ

LOOK FOR NORWAY MAPLES IN NOVEMBER

Norway Maples don't turn different colors—red, orange, yellow—as most other maples do. They turn as yellow in autumn as Sassafras trees, but they do it after most other trees are bare. If you couldn't tell them from Sugar Maples before, now you can. Look for Norway Maples in cities, where their tolerance of pollutants makes them especially popular. (See also page 205.)

NOVEMBER 8
Whistling Swan

This morning I glanced absently out the window and was startled to see seventeen large, white, long-necked birds floating about a hundred yards offshore. Was I seeing things? I called a nearby nature center and was told that Snow Geese and wild swans were the only large, white, migrating water birds along Lake Michigan. The birds' necks were too long for Snow Geese, which left the two native North American swans: the Whistling Swan (also called the Tundra Swan) and the Trumpeter Swan.

Some research helped solve the puzzle: In fall, Whistling Swans often appear in remarkable numbers, sometimes in the hundreds, along the Great Lakes, on their way from their summer home in the far northern tundra to the eastern seaboard wetlands where they winter by the thousands. (A similar number winter on the West Coast.) Since Trumpeter Swans are quite rare, my visitors were almost certainly Whistling Swans. The two are so similar that even if I'd seen them up close, I still wouldn't have known for sure. But I was able to eliminate Mute Swans absolutely.

I could hardly believe I had seventeen native, migrating swans within sight of my living room. I watched them for some time, hoping to see seventeen large white birds take wing—but they outsat me, and I went back to work. When I took a break later, the swans were gone.

(Afterword: A year later, hearing loud honking overhead, I ran outside with my binoculars, looking for geese, but discovered instead a hundred Whistling Swans, necks long as stretched arms, huge wings flapping gracefully, black legs tucked under their tails, black bills cutting the wind. It took several minutes for all of them to pass over. I don't think I shall ever again see anything so beautiful.)

№ № №

HOW TO TELL A NATIVE FROM A MUTE SWAN

You can tell a native swan from a Mute Swan, even from a distance, from the neck position and bill color. Distinguishing between the Whistling and the Trumpeter, slightly larger than the Whistling Swan, is harder.

SWAN	NECK POSITION AT REST	BILL COLOR
Mute Swan	Classic S curve	Orange
Whistling Swan	Fairly straight	Black
Trumpeter Swan	Straight, cricked back	Black

NOVEMBER 11

Bufflehead

Even though the wind seemed soft on this bright, blue morning, waves wrinkled the horizon. When I got to the beach stairs, I thought I saw a swan floating among a flock of ducks fairly far out. It was an optical illusion: Through the binoculars my "swan" became a Herring Gull. The gull, twice the size of the tiny ducks around it, appeared unrealistically large, and no wonder: The ducks were Buffleheads, the smallest species of waterfowl in North America, just half the size of the two-foot gull.

I counted them: eleven little, puffy-headed, black-and-white ducks. A dramatic white pie piece notched the top half of each male's green head (which looked black at a distance) while the brown-headed females showed a smaller white spot at four o'clock. I couldn't believe it! I'd been prowling around the local ponds and marshes since last spring trying to find a Bufflehead, and here they were, parked right in front of my house. Home is beginning to feel like a set for "Wild Kingdom."

As I walked up the beach, a few stray Buffleheads flew by, singly, in twos, in nines, skimming the whitecaps. Their wings beat very fast, their delicate black-and-white feather patterns intricate as butterfly wings. Soon soaring gulls appeared, close to shore, circling hungrily, hoping to steal fish from another group of thirty Buffleheads that bobbed in the waves like bathtub toys. Closer now, I watched them seem to play, upturning, diving, popping up seconds later, not far off. Pairs aggressively chased off male intruders, practically running on the water to do so. They were as comical as penguins.

🦆 🦆 🦆

SPOTTING MIGRATING DIVING DUCKS IN WATER

Watch for the gulls. Gulls love to steal fish, harassing not just fishing boats and ice fisherfolk, but fishing water birds, too. Although Buffleheads are startlingly white-bodied, their tiny size and dark wings can make them hard for beached observers to spot in large waves. The gulls have a vertical advantage, however, and often give the ducks' location away.

NOVEMBER 14

Goldeneye

 I'm getting better at finding ducks on dark water, staring until I see spots. They really are hard to see. This morning, my careful scrutiny was rewarded by fifteen Buffleheads, white polka dots in the winking whitecaps. Then, as I walked down the beach, eight larger ducks flew toward me, necks stretched out, wings whirring. Some of the ducks had brown heads with bright white neck rings. They landed in the waves, not far out, but soon dove deep and disappeared. I never saw them again. They were white-bodied diving ducks, possibly Goldeneyes. In the space of half an hour, about a hundred more black-and-white ducks flew by, low and fast. The Buffleheads were easy to recognize, but there were other white-bodied ducks I wasn't sure of. If the surf hadn't been so noisy, I might have known for sure.

🦆 🦆 🦆

THE WHISTLING DUCK

The easiest way to identify a Goldeneye is by the whistle of its wings as it flies by. If, however, crashing surf interferes, look at the female: The female goldeneye's head is brown, with a distinct, white neck ring, while the female Bufflehead's head is dark with a big, white spot at four o'clock.

Pileated Woodpecker

Another beautiful day: blue sky, bright sun, a nice little northwest tail wind. If I'd been a bird, I'd have done some serious migrating, so I went looking for Goldeneyes. I was walking toward the edge of one of my favorite backwaters, when I heard drumming in the woods. "It's just a Hairy Woodpecker," said my left brain with a shrug. Feeling a little silly, I decided to investigate anyway. Sneaking over the ankle-deep leaves, however, was like walking on cornflakes. I'd take a few steps, cringe at the explosion of sound, and the tapping would stop. I'd stop, and the tapping would resume. Step-step-step . . . tap-tap-tap . . . step-step-step . . . tap-tap-tap.

Soon, a second loud, hollow hammering began on my left, answering the tapping on the right. While I stood wondering which sound to pursue, a crow-sized, black bird floated in front of me, flashing a crimson crest and a great white patch under each wing. It landed about twenty feet away, right next to a similar bird. To my astonishment, not one, but two Pileated Woodpeckers had presented themselves. Although nearby I'd seen a whopping, straight-sided tree hole, such as Pileateds make, I'd never heard of anyone around here ever seeing one.

For twenty minutes, while squirrels crashed through the deep, dead leaves and November sunlight slipped through the complex weave of bare tree branches, I sat on a mossy fallen log, watching the magnificent pair work the trees around me. Often they landed so close that I didn't need binoculars to see the zebra stripes on their faces, their flaming red crests and long, powerful beaks. The female resembled the male, minus the front part of the crest and the male's red mustache.

Having learned that most woodpeckers are notoriously independent —most species are quite aggressive toward other woodpeckers, even (except during mating season) toward the opposite sex of their own kind—I wondered why the pair seemed so companionable. Later research revealed that although each Pileated Woodpecker sleeps in its own separate tree hole, unlike other woodpeckers, they hang around together from waking until they say good night.

In the southern states Pileated Woodpeckers are sometimes fairly common in rural and suburban areas, but in the North they shy from human eyes, hiding deep in woods offering the large, dead trees where Carpenter Ants feed. Guidebooks often warn of the difficulty in getting close to them. So why did they hang around me for so long? Maybe they just wanted me to give them an honorable mention.

🐦 🐦 🐦

THE PILEATED WOODPECKER: A DIFFERENT "DRUMMER"

There are two main kinds of woodpecker tapping: *hammering* at trees to get at insects or dig out nesting holes, and *drumming* on trees to communicate to a mate or a rival, or to establish territory. The drumming of the Pileated Woodpecker can often be distinguished by its louder volume and the tapering, soft way the drumming ends.

NOVEMBER 20

Robin's Nest

The leaves are all but gone. Only a few brown oak leaves interrupt the flow of branches against the sky. I'm always scanning the bare trees for life now, most of it having flown, and I often mistake a leaf on the wind for a bird. The loss of camouflage, however, has revealed evidence of quite a number of secret summer neighbors. A really large nest sits high in a Sassafras tree a stone's throw from my deck. An assortment of Easter egg–type nests are jammed into branch forks a little lower down, and across the street, tiny round baskets swing from branches like ornaments. A gray, papery hornet's nest, big as a melon, can now be seen swinging over the middle of the street, positioned perfectly to drop into a passing convertible.

It's empty now, of course, as are all the other nests, at least of the original architects.

Two weeks ago, I spotted a nest as big as two cupped hands and about five feet off the ground, in the middle of a twiggy shrub across from the public beach. This is not a private place: Joggers and strollers passed this bush, probably by the hundreds, on hot summer days. Still, no one knew the nest was there until the leaves went. I drove or walked past it about four times a day for a week, until today, when I showed it to Mary.

I thought it was a cardinal's nest like the one Mary had shown me outside her window last spring, but that nest was smaller and lighter. This nest was thick with grass and heavy with mud; it even had pebbles in it. "Oh, that's a robin's nest," said Mary. "There are even some shell bits left." The fragile shreds were robin's egg blue, the color of that band of sky that, just after sunset on a very clear day, lies between the last red and the dark.

&a &a &a

HOW TO RECOGNIZE A ROBIN'S NEST

Most birds don't use as much mud as robins do, or combine grass and mud to make such a heavy, bowl-shaped nest. Robins return early enough in spring to have plenty of water and mud to make nests. Also, robin's nests are lined, inside and out, with grasses only—no twigs or leaves.

NOVEMBER 25

House Sparrow

While Mary and I waited at the dinette for our corncakes to be served, we noticed a crowd of energetic, black-bibbed (male) and dull-breasted (female) House Sparrows enjoying breakfast at a birdfeeding station outside the window. The feeder's location next to the sidewalk didn't appear to disturb them at all.

WHEN IS A SPARROW NOT A SPARROW?

When it's a House Sparrow. House Sparrows, called "Old World" sparrows because they are not native to this continent, are actually a kind of weaver bird, no relation at all to our "New World" sparrows, which hail from the finch and bunting families.

"Why do you suppose House Sparrows like urban areas so much?" asked Mary. "I live only a mile from town, and I never see any of them." I didn't get House Sparrows at my feeders either, but that Mary didn't was far more remarkable. Every bird who's any bird makes an appearance at Mary's Seed 'n' Suet Buffet, substantial feeders stationed on three sides of her house.

In answer to Mary's question, I discovered that the House Sparrow, like the starling and the pigeon, which also have opted for urban life, is not native to North America. A number of House Sparrows (some sources say "a few," some say "fifty") were released 130 years ago in New York City. They became prosperous immigrants, flying blissfully into the American Dream, multiplying into the millions and leaving no state unturned. Like humans but unlike most birds, they have been known to reproduce in any season and boldly take advantage of urban offerings, which in earlier times included the seed-rich feed bags of carriage horses.

Despite its adroit thieving, the House Sparrow has not stolen the American heart. Introduced in some places for pest control, the sparrows became pests themselves, robbing crops (they also are attracted to agricultural areas) and chasing away local species. House Sparrows, like some other successfully introduced birds, are often aggressive.

Nature writers, who may wax eloquent over other common birds, are often unkind to the House Sparrow. Although the House Sparrow uses many calls to communicate, its song may be described as a "harsh chirp," its nest as "unkempt" and its appearance as "plain," even "ugly." I don't agree with that last bit; I find the black-bibbed male quite dashing.

ॐ ॐ ॐ

COMMON, EARTH-TONED, BLACK-BIBBED BIRDS

If you are confused by small, black-bibbed birds, look first at the beak: A sparrowlike bird with a black bib and a black beak is a House Sparrow, unless it has a black cap, too, in which case, it's a Chickadee. Black-throated birds not mentioned here are usually bigger or brighter than these sparrow, or sparrowlike, species.

NOVEMBER 28

Mints

The woods have stilled. The air feels thick as cold broth. The towering trunks I walk through sometimes groan like animals; otherwise, silence. I get impatient with this lack of life. I am accustomed to new things blooming, growing, flying, crowing. Spring, summer and fall offer daily surprises, like courses at a Chinese banquet. November fare is simpler, plainer, brown, like meat, potatoes and gravy. I can't yet appreciate its subtleties.

Today I glanced through the bare branches of trees, looking for movement. I skimmed the brown ground for green. There's still some left: defiantly emerald-green moss, the less intense Evergreen Woodferns (see page 31), gray-green lichen curling delicately from dead or alive wood. Finally, I found a tight mass of tiny, tender leaves sprouting from a plate-sized patch of ground. When I picked some, an intoxicating scent exploded into the air. I took a handful of leaves home and, even without flowers, soon placed them in the mint family.

Mint is a very large family, including aromatic superstars like Oregano, Marjoram, Basil, Thyme, Sage, Spearmint, Peppermint, Catnip, Pennyroyal and even Lavender. If I'd had a flower or even a full-grown plant, I might have known which mint it was. But a mint it was for sure. Members of the mint family, even though many of them are imports from Europe, share some distinctive characteristics.

٭ ٭ ٭

HOW TO IDENTIFY A MINT

If a plant has opposite leaves on a square stem, there's a very good chance it's a member of the mint family. If it's aromatic, you're even closer: Many (not all) mints are deliciously scented. Clusters of tiny, lipped, lavender or purplish flowers—although some mints flower white, red, blue, even yellow, most run to shades of lilac—bring you almost home. Mints usually blossom in clusters along the stem or at the top.

NOVEMBER 29

Wild Thyme

I called the aromatic plant I found yesterday Wild Thyme, based, quite unscientifically, on a sniff test conducted in front of my spice cabinet. Of all the herbs that resembled it, including Basil, Marjoram and Sage, Thyme came closest. I hoped this was correct, because Wild Thyme flowers boast a curious characteristic not shared by all mints: The flowers are hermaphrodites, blooming the first day as males—sometimes shedding pollen, sometimes not, depending on the variety—and the next day as female.

It's not easy to tell members of the mint family apart; there are so many of them (thirty-two hundred species worldwide), and they look so much alike. Wild Thyme is one of the easier mints to identify; it's a creeper, growing in dense, vinelike cushions, as few mints do. Scent sometimes helps narrow possibilities, but flower color, in warmer seasons, can help, too.

WATCH OUT FOR STINGING NETTLES!

I haven't found any poisonous mints, but Stinging Nettle, although not a mint, does resemble one, having a four-sided stem and opposite leaves. Stinging Nettle stings painfully when touched, so if the stems and leaves of a mintlike plant are bristling with hairs, identify it before touching it.

᭰ ᭰ ᭰

WHICH MINT?

Most mints have opposite leaves, a square stem and clusters of tiny, two-lipped flowers that blossom either closely along the stem or in a terminal cluster at the top (there are exceptions). So how do you tell what's what? Flower color helps narrow the wild number of possibilities.

FLOWER COLOR	*MINT SPECIES*	
Yellow	Horse Balm	European Wood Sage
White	Wild Mint Water Horehound	Hoary Mountain Mint
Red	Bee Balm (Oswego Tea)	Salvia (Scarlet Sage)
Blue	Blue Salvia American Pennyroyal	Blue Curls
Blue-violet	Ground Ivy Hyssop Skullcap	Lyre-leaved Sage
Purple	Wild Thyme	Self-Heal
Pink	Wild Marjoram	Obedient Plant
Pink-lavender	Wood Sage (Germander)	Wild Basil
Lavender	Hoary Mountain Mint Henbit Peppermint Wild Bergamot Catnip	Wild Mint Motherwort Spearmint Plains Bee Balm

December

DECEMBER 2

Mars

At five o'clock this morning, moonlight splashed in my face and woke me up. Outside my window, it poured into the lake, lighting up ruffled waves, silver on black. Dripping from the moon's east side was a strangely bright planet or star, I didn't know which. It was a scene to make even a bleary person gasp. I got up, dressed warmly and went down to sit on the top step of the wooden beach stairs, sip hot coffee and take in the show. The brilliant moonlight washed out most of the stars, but three were so bright that I was sure they were planets. I knew one of these was Mars. "StarDate," which I hear on the University of Michigan National Public Radio station, has been reminding me that during these few weeks, Mars is closer to Earth than it will be again until after the year 2000.

Last summer, when I met an astronomer who could read the stars like Braille, I became aware of how little I "saw" in the sky. When he looked at the sky, even a moonless sky milky with stars, the pricks of light were as sensible to him as the tracks of letters on a page. All I saw was ambiance. I could only name the Big Dipper, the Little Dipper and Orion. I felt as I did when I began naming nature, when I looked outside and all I knew were Cardinals, jays and crows.

Astronomy is a subject for another book, but I might be excused for mentioning the universe when it makes so brilliant an appearance over my front yard. I called the astronomer, who reported stargazing this morning as well, and this is what he told me: The planet closest to the moon was Mars. The brightest one, the one at eight o'clock, was Jupiter. The light at eleven o'clock was probably Sirius, the "Dog Star."

❧ ❧ ❧

HOW TO TELL PLANETS FROM STARS

I've found no sure trick to help the beginner pick out the planets, but there are hints for earnest skygazers.

Planets change position in relation to the stars, stars don't. However, it takes observation over many nights to tell what's moving and what's not.

Stars twinkle and planets don't, according to folk wisdom. This may be true sometimes, but not always. "Twinkle" is the lay term for "scintillate."

Look for the *ecliptic*. Because the planets circle the sun, they follow close to the sun's path through the stars, a path called the ecliptic. If you can find a couple of bright planets, other planets present will nearly line up with them. Planets will often appear to be in an almost straight line near the ecliptic.

THE BRIGHTEST NIGHT LIGHTS

Our brightest planets are usually Venus, Mars and Jupiter. Our brightest stars are often Sirius, Arcturus, Vega and Capella.

DECEMBER 5

Red-bellied Woodpecker

Although I work facing a wall, I can still observe the woods and deck reflected in the glass of a painting that hangs over my computer. Today I saw in this double exposure a Red-bellied Woodpecker clinging to the suet on the deck as well as to the torch in the painting's Statue of Liberty. Its silhouette was so much sleeker than the usual Downies and Hairies that I was immediately alerted.

I thought I'd seen all the neighborhood species of woodpeckers, but I hadn't seen a woodpecker like this one. A wide ribbon of red ran across the top of its head, from its beak to its back, which was zebra-striped, unlike those of our other local woodpeckers. The breast looked like sheen-sanded, lion-hued wood. I don't know why it's called a Red-bellied Woodpecker, because the rosy stain on its belly is difficult to see and is certainly not its most obvious feature. I had to look through several guidebooks to find a picture that showed it; until then, I was convinced that the Red-bellied Woodpecker wasn't red-bellied at all.

❧ ❧ ❧

WOODPECKERS WITH "LADDER BACKS"

If you see a ladder-backed woodpecker—one with black-and-white stripes on its back as well as on its wings—in the eastern United States, a Red-bellied Woodpecker is a very safe bet. There are other ladder-backed woodpeckers, but most of them—the Golden-fronted Woodpecker, the Gila Woodpecker, the Ladder-backed Woodpecker and the Red-cockaded Woodpecker—are found in the South. Nuttall's Woodpecker is found on the West Coast, and the Three-toed Woodpecker in the Rockies and Canada, far from the Red-bellied's range.

WHERE THE RED IS ON COMMON EASTERN WOODPECKERS

There are a surprising number of woodpeckers, most of them black, white and red. The pattern of red, usually on the head, is so distinct that it's a good clue to identification.

Hairy Woodpecker (male):	Splash of red on back of neck
Downy Woodpecker (male):	Same as above
Common Flicker:	Red crescent on back of head
Red-bellied Woodpecker (female):	Ribbon of red on back of head
Red-bellied Woodpecker (male):	Ribbon of red from bill to neck, on top of head
Pileated Woodpecker (female):	Large red crest
Pileated Woodpecker (male):	Same as female plus "mustache"
Red-headed Woodpecker:	Head entirely red, like a hood

DECEMBER 11

Christmas Trees

It was still pretty early to dive into Christmas, but this brisk day begged for an outing. Two friends and I took a ten-minute drive to a local Christmas tree farm owned by a couple who, now in their seventies, run it as a retirement business. "We shipped out ten thousand trees

---

WHERE DO CHRISTMAS TREES COME FROM?

According to the National Christmas Tree Association, nearly 100 percent of our Christmas trees are commercially grown, 90 percent on U.S. farms and 10 percent on Canadian farms. Some state forests issue permits to individuals for cutting trees in selected areas, but forest cutting accounts for a very small percentage of Christmas trees.

---

this year," Howard told us, pointing to a pile of evergreens, each snug in its hair net. "That's the last of them."

Ten thousand trees? It didn't sound like retirement to me. "We have 150 acres of trees," Sally told me. "We plant fifteen hundred trees per acre, but we can only harvest about half of those." Christmas trees don't grow like cabbages, either: In this climate, a three-year-old Douglas Fir seedling takes ten years to reach living-room size, a Scotch Pine, half that time.

We were offered a choice: Scotch Pine, Colorado Spruce, White Pine, White Fir or Douglas Fir. We drove through the Douglas Firs and stopped amid a lush selection to wander with a small saw. I found my tree right away, in plain sight of the car. Of the most available Christmas tree choices, my favorites are Balsam Fir and Douglas Fir. I find the long pine needles tend to obscure my lifetime collection of ornaments and sharp spruce needles irritate my hands. Fir needles, however, are short and soft, and they smell nice, too.

ِ‌ِ‌ِ‌ِ‌ِ‌ِ‌

WHAT'S THE BEST CHRISTMAS TREE?

According to the National Christmas Tree Association, the bushy Scotch Pine leads the pack, with about a third of the market. The soft-needled Douglas Fir comes in second, with a market fifth. The remaining half offers a wide variety of spruces, pines and firs. Christmas trees can be a very personal or regional matter. The aromatic Balsam Fir, for example, is popular in its native New England, the soft White Pine in Michigan; the beautiful blue Colorado Spruce is a frequent Rocky Mountain choice, while the greenly graceful Monterey Pine is a California favorite.

HOW DO YOU KEEP A CHRISTMAS TREE FRESH?

I have encountered a number of concoctions designed to encourage Christmas trees to hold onto their needles instead of installing them in the carpet. One mixture combined corn syrup, plant food and chlorine bleach with water. Another suggested pouring a can of citrus soda in the water. The National Christmas Tree representative didn't think much of those, suggesting the following:

1. Buy a fresh tree.
2. Cut a slice off the bottom, much as you do cut flowers.
3. Keep in water *all the time*; don't let the water run dry.

DECEMBER 13

Douglas Fir

I have just discovered that no firs grow in this part of Michigan—or in most of the United States, for that matter—which explains why I haven't seen any here. I'd have to travel north or to the Rocky Mountains to find most of them in their native environments. At Christmas, how-ever, they can often be found at tree lots, or in my case, the Christmas tree farm. Now I have a Douglas Fir exhibiting all sorts of Douglas Fir char-acteristics right here in my toasty living room.

Actually, treating a Douglas Fir as a fir is cheating. In the early 1800s, a sample of the Douglas Fir was brought to the Royal Horticultural Society of London by David Douglas, an adventurous Scots naturalist who risked his life numerous times (and eventually lost it at the age of thirty-five) collecting North American plant species. Hoopla about what to call the odd evergreen went on for years. This was the controversy:

Douglas's fir had soft, short, flat needles that grew singly along the branch just as they do on a fir, but it couldn't be a true fir, because true fir cones grow upright, like candles. Although Douglas Fir cones dangled like spruce cones, it wasn't a spruce either, because spruce needles are squarish and sharp, not soft and flat. It couldn't be a pine, because pine needles grow in bundles. So what was it? Botanists tried to call it a "Douglas Tree," but "Douglas Fir" stuck, and is now officially a family as distinct as pine, fir and spruce. It's so distinctive that it's easy to tell it from other conifers.

᷉ᷓ ᷉ᷓ ᷉ᷓ

EVERGREEN CONES

If a conifer is bearing cones, you're in luck. Where and how the cones hang is as helpful for identification purposes as their appearance.

> Douglas Fir cones are papery, with distinctive three-forked "dragon tongues" extending from each "petal." The cones hang along the branch.
> Fir cones are papery, sitting upright, like candles, along the branch.
> Spruce cones are papery and hang along or at the end of the branch.
> Pine cones are woody and hang variably, depending on species.
> Hemlock cones are tiny, woody and drip from the ends of branches.

EVERGREEN NEEDLES

If it's the wrong time of year or the tree is too young, no cone is in sight. A Douglas Fir, for example, enters puberty about the same time humans do, at about twelve or thirteen years old. You can still tell many evergreens apart by the needles.

> Fir needles are short, soft, and flat; branch ends are rounded.
> Douglas Fir needles resemble fir needles, but the growing end of the branch is pointed.
> Spruce needles are sharp and squarish.
> Pine needles grow in bundles and are usually longer than fir or spruce needles.
> Hemlock needles are soft and flat, but show two light lines underneath.

DECEMBER 14
Geminids

"Be sure you don't miss the Geminid Meteor Shower during the early morning hours of December 14th," my astronomer friend called to tell me. "It's the best meteor shower of the year." I did need some urging; it certainly had never occurred to me to rouse myself from sleep to look at a sky that presumably is there all the time. Furthermore, expecting a clear sky along the lakeshore in December is like asking for snow in July. But the universe was not going to miss a chance to feature in my book: At 1:30 A.M., I awoke to a cold but cloudless, moonless, windless, *starry* night.

I pulled on long underwear, two sweaters, knit pants, two coats, two scarves, a pair of gloves and a hat. Then, flashlight in one hand and tea mug in the other, I stumbled through the dark to a beach that glistened in the starlight. Lying in sand as cold and hard as ice, wrapped in a quilt and sipping hot mint tea, I began scanning for meteors.

At first, all I saw were shooting stars. They fell far away, faintly and very fast. As my tea cooled, however, I saw the Geminids. They took my breath away. The Geminids fell near, slowly, glowing brightly. Some left tails like comets across the sky. Some fell south, some north, some west. Sometimes there'd be one, and then I'd wait five minutes and five would fall in a space of seconds. I spent an hour alone on a sandy beach that stretched to dark horizons, watching the starworks, gasping, shouting and cheering.

I was surprised to discover that meteors are comet garbage: As a comet moves around the sun, it leaves a trail of trash that spreads out like a jet trail. Most of it is smaller than a pinhead and burns up due to friction in the upper atmosphere. When the earth passes through a comet trail, we see a meteor shower, which is why astronomers can predict when meteor showers will happen.

❧ ❧ ❧

WHEN TO WATCH FOR METEOR SHOWERS

Meteor showers are fairly predictable and are named for the constellation from which they appear to "fall." Geminids appear to radiate from the constellation Gemini (northeast of Orion, which I find easier to spot). The following dates are not exact. Your local newspaper may announce a more specific time, as well as viewing conditions. Usually the show is best after midnight.

January 4:	Quadrantids
August 12:	Perseids
October 21:	Orionids
December 14:	Geminids

DECEMBER 20

Earthshine

I was changing the sheets on a guest bed early last evening, anticipating a visit from my college-age son, when I happened to look up and catch a brilliant crescent moon framed in the small window, which appeared to hang on the bedroom wall like a glossy museum poster. The round of moon glowed ethereally, cradled in a brighter, silvery sling, an effect sometimes called "the old moon in the new moon's arms." An hour later, I read a newspaper column which suggested that it was a very good night for observing *earthshine*. The term was not defined. I'd heard of sunshine and moonshine, but I'd never heard of earthshine. I was on the phone immediately to my astronomer friend. Here's what I learned:

Just as sunshine reflects off the moon to the earth, it also reflects off the earth to the moon. The moon's phases and the earth's phases are complementary (i.e. opposite), so when the moon is full, the earth, as seen from the moon, is a crescent. When the moon is a crescent, the earth would be seen as full from the moon. Earthshine is so bright—thirteen times brighter than moonlight—that it makes the dark part of the crescent moon visible, reflecting all the way back to Earth. Earthshine can appear blue, because the Earth is blue, thanks to our many oceans. Is this what is known as a "Blue Moon"? (See page 309.)

🐦 🐦 🐦

SKY NEWS

A very nice sky-map calendar is available that alerts you not only to moon activity but to all sorts of drama unfolding silently above your head. The maps are simple and clear, showing what to look for above the horizon in a selected direction at a selected time. Write to *Sky Calendar*, Abrams Planetarium of Michigan State University, East Lansing, MI 48824. (At this writing, the price was $6.)

DECEMBER 23

Climbing Bittersweet

I've recently seen two gorgeous wreaths made of entwined, woody Climbing Bittersweet vines. Both wreaths hung in the homes of dedicated environmentalists. Did I have the heart to tell them that Bittersweet is on the state of Michigan's list of protected plants? Well, yes, I did. That list used to be called "The Christmas Tree list"—and still includes evergreens, Michigan Holly and Mistletoe. In Michigan, one needs written permission to take from someone else's property any part of any plant on the protected species list. Some beautiful winter plants have been cut and collected so drastically that they could be headed for the even more tragic endangered species list.

I'd never heard of Bittersweet before this year, but of all the Christmas plants, it's the one I have the hardest time leaving alone. It is beautiful. The small, bright orange, cherry pit–sized fruits split at maturity, like orange-peel petals, to reveal a berrylike, crimson interior. Bittersweet was once fairly common in the Midwest and East, from southern Manitoba and Quebec as far south as Louisiana. Its berries stay brilliant for a long time and probably won't be confused with anything else. Nightshade (see page 241) is sometimes called Bittersweet, but its berries are red. Both fruits are poisonous.

CHRISTMAS DAY

Carolina Wren

Family visits, holiday festivities and freezing weather have cooled my ardor for nature walks, but today nature came to me, in the middle of Christmas dinner. My son, some friends and I were visiting with Mary and her family when her daughter, Joan, suddenly exclaimed over a pair of fat, brown Carolina Wrens swinging from a hunk of suet outside the French doors. A hushed uproar ensued around the table as all of us craned our necks to get a glimpse without scaring the birds away.

Carolina Wrens were an unexpected gift. The largest of all eastern wrens, they are as big as small sparrows, rich, clay-ruddy on top and buff-apricot beneath, with a bright slash of white over each eye. They went at the suet energetically with thin, beautifully curved beaks; insect-eaters most of the year, Carolina Wrens will settle for seeds and suet in winter. As with other wrens, their tails shot straight up, flicking excitedly.

I owed Mary an apology. Over a month ago, she'd told me about seeing Carolina Wrens at her feeders, but I never saw them myself and I didn't know whether to believe her. (Mary doesn't give two biscuits about naming anything, no matter how passionately I wax about the subject.) I doubted that even Mary's deluxe feeders would tempt the state bird of South Carolina this far north. As it happens, however, Carolina Wrens are known to spend milder winters here, if infrequently.

Joan sounded eager to get the Audubon Society's rare bird alert "phone fan" going, a system in which one person calls several members, each of whom call several more, and so on. Soon out-of-town birders will be crunching in lakeshore snow, hoping to add the Carolina Wren to their Life Lists. (Afterword: Sure enough, they did.)

ào   ào   ào

WRENS WITH WHITE EYE STREAKS

There doesn't seem to be one thing to look for to tell the three wrens that resemble each other—the big, ruddy-brown Carolina Wren, the sleeker, grayer Bewick's Wren and the tiny Marsh Wren—apart. When all wrens stop looking alike to you, look first for the white eye streak and then for the following:

WREN	RANGE	FIELD MARKS
Carolina Wren	Eastern United States	Buff, orangish breast; plain, brown back
Bewick's Wren	South, Midwest, West Coast	White breast and tail edges
Marsh Wren	Summers North, winters South	White breast with brown flanks; streaked back.

DECEMBER 27

Bald Eagle

Bald Eagles were big news at last night's party; and this morning, Mary called to say she'd seen one hunting along our beach, and Joan had seen three along the Kalamazoo River, just minutes from here. This report came on the heels of a radio announcement of this week's Annual Eagle Count, sponsored by the Michigan Department of Natural Resources, together with the National Wildlife Federation. There's been a dramatic rise in eagles sighted in Michigan: One year the count was as low as 32, but 359 were counted last year, up from 200 the year before. (I wonder if there are more eagles, or just more people counting them!)

Today I went out to find an eagle for myself, there being just a few days left to put eagles in my book. I never really thought I'd see one, crunching slowly down a snowy River Road in my van, scanning the water through the bare trees. In less than ten minutes, though, there it was: on an island in the middle of the river, a huge dark knob resting on the spindly top of a tall, bare tree. Through my binoculars, the image was unmistakable: big dark body, white head, white tail and hooked, yellow beak.

Of our two largest eagles—the Bald and the Golden—Bald Eagles are the most commonly seen in the eastern United States. Unlike the Golden Eagle, the Bald Eagle is really a sea eagle, a fish-eater that nests near water. Although it prefers fish, it will eat meat, too, dead or alive, and is sometimes spotted along highways, feeding on road kill. The Bald Eagle is also a notorious thief, harassing the smaller sea eagle, the Osprey, which is a much better fisherbird. The Golden Eagle is more common in the West, where it is often seen hunting small mammals or birds in the mountains and grasslands.

Bald and Golden Eagles are usually so large, dark-winged and dark-bodied that they are easily distinguished from hawks and Turkey Vultures. But how do we tell eagles from each other?

Identification problems come with young eagles, which can take up to five years to reach adult plumage. A juvenile or subadult (as those in the early developmental stages are called) Bald Eagle closely resembles an adult Golden Eagle, except that it has no golden neck sheen. Location can sometimes help identify a young eagle. One seen in the eastern United States near water, fishing, or eating fish is most likely a Bald Eagle. Although Bald Eagles are not uncommon in the West, an eagle hunting small mammals in the mountains and plains is likely a Golden Eagle.

Afterword: About a week later, an immature Bald Eagle cruised so near my house that I could see the white mottling on the dark underwings, look right into that eagle eye. The huge, dark bird flew slowly in from the lake, through the pristine, snow-filled woods and right past my windows, before vanishing over the trees.

ᏧᎠ ᏧᎠ ᏧᎠ

BALD OR GOLDEN EAGLE?

The adult Bald Eagle, which is not bald at all, has a white head and white tail, while the adult Golden Eagle is mostly dark brown, with a golden sheen on the nape of the neck.

ADULT BALD EAGLE	*ADULT GOLDEN EAGLE*
SIMILARITIES	
Big brown body (27 to 35 inches)	Big brown body (27 to 33 inches)
No white under wings	No white under wings
Huge wingspread (71 to 89 inches)	Huge wingspread (72 to 82 inches)
DIFFERENCES	
White head and white tail	Golden sheen on nape of neck
Range: East and West	Range: mostly West and Alaska
Soars flat	Soars in slight dihedral
Prefers fish, but will eat meat	Eats meat
The national bird	

DECEMBER 29

Rough-legged Hawk

The river in winter, I discovered, is a great place to look for hawks. I didn't just see an eagle yesterday; I also saw four hawks inside of half an hour. They were easy to spot through the leafless trees: big lumps perched on thin branches along the iced edge of the water. Three of them spooked when I stopped my van, but a fourth flew, hunting, over the river for a very long time, wings slowly lifting and lowering, like a Sting Ray in deep water, wheeling slowly around a small island that bristled with brush. Sometimes it hovered for as long as twenty seconds. I thought I was seeing things; the only other birds I'd ever seen hover were terns and hummingbirds. It also looked unlike any hawk I'd ever seen: white, mottled body and underwings, with a dramatic, square black patch at the joint of each wing. The tail was white, ending in a wide, black band.

No hawk in my guidebook had a tail like that—or big black spots on the wings, for that matter—except the white morph of the female Rough-legged Hawk. The Rough-legged Hawk has a light and a dark form, but both retain the distinctive dark patches on the wings. The male has more stripes on its tail. As I read further, my identification was confirmed: The Rough-legged Hawk is known for its unusual hovering. Named for the feathers that grow to its feet, it is an arctic hawk that winters all over the United States.

NEW YEAR'S EVE
Blue Moon

Once in a Blue Moon, a full moon comes around twice in the same calendar month. When this happens, the second full moon is called a Blue Moon. The expression "Once in a Blue Moon" means "not very often"; two full moons occur in one month only seven times every nineteen years. On this last night of this particular year and just in time to make the last entry in my naming nature journal, a Blue Moon will appear, sailing high in the sky, as winter full moons do, into the morning of the New Year. (Summer moons ride low in the sky, where the atmosphere makes them appear particularly golden. Winter moons ride high, giving them a lighter, silvery look.)

I got a preview last night, when the clouds parted briefly and out swung a bright white moon. It didn't look blue at all, and it won't tonight, either. A blue moon never does, unless the earth's atmosphere is full of debris from a volcanic eruption or some other dramatic event. Actually, a faintly blue moon isn't rare: Sometimes, in the early crescent stage of the moon, the usually dark part glows faintly blue from sunlight bouncing off the Earth's watery surface (see page 302).

I love ending this year, and my book, with the Moon. The Moon has long been a symbol for feminine power, knowledge and mystery. I am feeling that power lately, along with a similar magnetic attraction to the Earth I never felt before this year. Naming nature has brought *landscape* into sharp focus, as if my eyes have been fitted with some new technological device. Outside my window now are not just birds, as last year at this time, but nuthatches, chickadees, Red-bellied Woodpeckers, House Finches, titmice, Downies and Hairies. Going after sunflower seeds are Gray and Fox Squirrels. I look out at the boughs not just of evergreens, but of Eastern Hemlocks, Red Cedars, White Pines, Norway Spruce and a hedge of yew. The bare branches I know, too: White Ash, Basswood, American Beech, Sugar Maple, Sassafras, honeysuckle and Osier Dogwood. Still green and visible under last night's powdering of snow is not just ground cover, but Myrtle, Ivy and Pacasandra. Evergreen Ferns fountain along the bank.

Tonight I will name the Moon, and even the planet two degrees from it. But what brings me joy is more than knowing names. Many names, like last summer's wildflowers, have already faded away. It was the *process of naming* that drew me like a magnet into a magnificence of green, pulled me into flowers, opened my ears to the cadences of birds. For the first time, I really feel like taking care of things. As with any love, naming nature has resulted, for me, in paradoxical emotions: The joy of discovery is accompanied by an acute fear of loss. I have never in my life made a more heartfelt New Year's toast than this one: *Happy New Year*, Earth! Long may you be named, nurtured and loved.

ﺰ ﺰ ﺰ

NAMING THE FULL MOONS

Every full moon has a name, sometimes several, and occurs in sequence, usually, but not always, in the month listed below. Many of the names are ancient, translations of certain Native American moon names, but they are still used today.

January:	Old Moon or Moon After Yule
February:	Snow Moon, Hunger Moon or Wolf Moon
March:	Sap Moon, Crow Moon or Lenten Moon
April:	Grass Moon or Egg Moon
May:	Planting Moon or Milk Moon
June:	Rose Moon, Flower Moon or Strawberry Moon
July:	Thunder Moon or Hay Moon
August:	Green Corn Moon or Grain Moon
September:	Fruit Moon or Harvest Moon*
October:	Harvest Moon*
November:	Hunter's Moon or Beaver Moon
December:	Moon Before Yule or Long Night Moon

* The Harvest Moon is the full moon closest to the equinox, which occurs in late September or early to mid-October. There may also be a Honey Moon, which is usually a June moon, named for its color.

RING AROUND THE MOON

When the Blue Moon rose tonight, it was circled by a glowing pink ring. Rings around the sun and moon are called *haloes*, caused by light refracting through ice crystals in the earth's atmosphere, in much the same way that a rainbow, another ring, is caused by light refracting through droplets of water.

Afterword

I have been blessed with an editor, Mindy Werner, who just won't leave loose ends alone. After reading my completed manuscript, she wondered how I felt now that my naming nature project was over. Beyond making me environmentally aware, had it changed me personally in any significant way? Had it helped me in the healing process during that first year after my divorce? Mindy suggested that I add a postscript. So here it is:

It has been just over two years since I began naming nature. The first year I kept a journal; the second I spent researching and writing this book. What began as a personal project grew into a book proposal, then became a way of life. My days began dancing to the rhythm of my work. I made countless new connections, not only with nature, but with friends, who frequently contributed their knowledge and companionship. My world expanded. At one time quite sedentary, I began walking, becoming intimately acquainted with the beaches, meadows, orchards, lagoons, woods, rivers, lakes, ponds, golf courses, wetlands and rural roads around me. I was even getting some exercise. I was frequently overcome by joy. New connections with nature, new friends, new connections with old friends, an expanding environment, physical fitness and joy do much to heal a heart. They don't fix it, but they support the process. They nurture growth.

Naming nature, however, was part of a much larger personal project, which was to learn to recognize, honor and nurture the feminine in myself, in others (both men and women) and in the world. I've been trying to balance what I think of as my masculine side (the side that values critical thinking, reason, analyzing, organizing and utilizing), which has long dominated my life, with my feminine side (the part of me that values intuition, feeling, playing, nurturing and protecting). To this end I have also worked with dreams, women's groups and therapists. But connecting with nature has been an important part of it all, a way to go in by going out, to simultaneously work and play, to make the masculine serve the feminine for a change. It has kept me from living too much in my head or falling into my navel, inspiring the close cooperation of body, heart and mind, an unusually effective committee.

Nature as ambiance felt too feminine to me; nature as resource (fuel, food and fun) too masculine. Naming (masculine) nature (feminine) used both sides of myself, helped put my life and work in balance. Some people have a problem with using masculine and feminine in this way, but this is how I see myself. For me it works.

Saying good-bye to this book is difficult. It has been as much a part of my life as a lover. But I have other books to write, dreams to realize, passions to explore, and I just can't wait.

—M. B.

A Guide to Guides

Nature guides are much like sparrows: At first encounter, they all look pretty much alike, with more species than seem reasonable. Despite this bewildering first impression, however, nature guides are not the same. In the course of writing this book, I purchased many of those commonly available, and although most were useful in learning more about what I'd seen, I found only a few that met my outdoor identification needs. If I'd known what to look for, I'd have added to my outdoor pleasure and avoided some frustrations.

If you want to buy only one guide for a subject, you will have to decide which is more important to you: the amount of information given on each species or the ease of identification. It's difficult to have both in the same book: Copious information makes a guide bulky, but if you don't need to take it into the field, it won't matter. My favorite tree guide, for example, is oversized, but I often bring the leaves and twigs home. A bird guide, however, should, in my opinion, be quick and easy to use. You can't take birds home with you, nor are they apt to stay put for much paging.

Below are the qualities I found myself looking for in guides I wanted to accompany me outdoors. The price for these conveniences, of course, is often a lack of descriptive detail.

When I began using nature guides, I found myself asking several questions that didn't seem to have ready answers. For example, just how inclusive were these guides? Did a bird guide have *all* the bird species in it? As I used the various guides, I found that, generally speaking, the major guides to North American birds and animals were most complete—it was

---

HOW TO CHOOSE A FIELD GUIDE

If you want only one field guide for a subject, be it trees, birds or wildflowers, look for the following advantages:

Portability: A guide should slip into a back pants pocket.

One-stop organization: It's really helpful to find everything on one species —picture, information and range map—in one place. This is especially important for bird guides, with which time is of the essence.

Clear art: I like drawn, color art for identification—special markings tend to be clearer—but some people prefer good photographs.

Home range: Check the title page for the area covered by the guide. I've found several guides with misleading covers: A guide I assumed covered North America actually included the entire world; another, this one a guide to the Northeast, appeared at first glance to cover the continent.

---

rare to find a species not listed—but there are too many plants, insects and mushrooms to allow for a manageable, comprehensive guide. I also found that guides covering the widest ranges, such as the world, as well as those about the narrowest, such as guides to species in one state, seemed the least inclusive, the first because of the overwhelming number of species to consider and the second probably for lack of funds. I settled on reputable guides that covered either North America or my section (Northeast) or half (East) of the continent. (See also comments on individual books.)

Another question: How many guides does a nature namer need? It's not necessary to buy a guide on every subject to enjoy naming nature. A bird and a wildflower book would probably do fine for starters, unless one has special interests. It's not possible for the following list to include all relevant books published—I've only reviewed those I've used—but the principles can be applied to anything currently available. Specifics on authors, publisher and date of the edition I used can be found in the Bibliography. Prices are included only as a general guide; by now, they're probably higher. The arrangement is quite personal: My favorites come first.

A Guide to Almost Everything

Reader's Digest North American Wildlife. If I were to buy only one guide, I would buy this one, even though it's hardcover and too heavy to carry outdoors. The full-color illustrations, however, are clear and beautiful, and a little range map accompanies every one of the two thousand plants and animals included. This is especially useful for plants, the ranges of which are usually described verbally. Although it makes no attempt to be inclusive, most of the subjects in *Naming Nature* are found here. The wildflower section is particularly strong. Other subjects include birds, reptiles, trees and shrubs, mammals, fish, mollusks, insects, ferns, mosses, seaweeds, lichens and mushrooms.

Some Guides to Birds

Birds of North America: A Guide to Field Identification (Golden Field Guides). This book meets all the necessary qualifications for a first-rate field guide: It slips easily in and out of a back jeans pocket, puts all information on a bird in one place and uses clear (if not magnificent) illustrations. I didn't mind that many species outside my range were included; the range maps were right there. This book is for identification only; little information is included. ($11.95 paper)

   Eastern Birds (Peterson Field Guides). I found that these excellent guides are preferred by many birders, combining Peterson's fine illustrations with more description than is available in Golden Guides, but I felt it fit in my back pocket too snugly, and I was frustrated by having to look up the range maps in the back of the book. (*Western Birds* is also available.) ($14.95 paper)

   Field Guide to the Birds of North America (National Geographic Society). This beautifully illustrated guide would be the perfect companion for outdoor excursions if it weren't so big and heavy. It would fit in a jacket pocket, though, and does put all its excellent, well-written information in one place. A wonderful companion to a lighter field guide. (Available through the National Geographic Society.)

   The Audubon Society Field Guide to North American Birds: Eastern Region. The thickness and two-stop information placement of this volume made it awkward for me to use in the field, and I didn't find the photographs too helpful, either. The write-ups on each species, however, are excellent and often extremely interesting. Since I bought my guide, a new edition has been published which deals with some of these problems and would be worth considering. (*Western Region* is also available.) ($17.95 paper)

The Audubon Society Master Guide to Birding. Mostly photographs of startling quality, picturing the birds of North America, with expanded information on each species. This heavy three-volume set, though not for easy toting, is a paradise for armchair bird-watchers. ($47.85 per set paper)

Some Guides to Wildflowers

Newcomb's Wildflower Guide. Although it doesn't say so on the cover, this guide is limited to the northeast quarter of the United States, Ontario and Quebec, a fact I wasn't aware of until I'd used it for almost six months. It doesn't use much color, either; the arrangement is based on the number of petals, or parts, on a flower. This organization took some getting used to, but I liked it because, within its range, there just wasn't anything I couldn't find. I found this preference shared by several botanists teaching the subject. ($14.95 paper)

A Field Guide to Wildflowers: Northeastern/Northcentral North America (Peterson Field Guides). Light, compact and pocket-sized, this guide is arranged by flower color, which makes identification simpler for the novice, and it seems reasonably inclusive. (Other regions also available.) ($14.95 paper)

Michigan Wildflowers in Color, by Harry C. Lund. I include this as an example of a really good local guide. The color photographs aren't always clear, there are a number of flowers I couldn't find here, even Michigan flowers, and the book was too big for my pocket, but I still carried it whenever possible. The comments were more extensive than I found in the above two books, and it was the only wildflower book I found that included information on a flower's endangered status. Local guides like this one have their advantages and are often less intimidating to use than more inclusive guides. ($14.95 paper)

The Audubon Society Field Guide to North American Wildflowers: Eastern Region. I had the same problems with this guide as I did with the Audubon bird guide: too fat; the photographs, though quite clear, did not always show the leaves; and the information was printed separately from the picture. However, although I didn't use it as a field guide, the information included was far more extensive than any of my other guides, and

I wouldn't be without this book. (*Western Region* also available.) ($17.95 paper)

Some Guides to Trees

Trees of North America: A Guide to Field Identification (Golden Field Guides). This guide claims to cover "594 of some 865 species of trees native to North America north of Mexico." Small, light and reasonably inclusive, this was my preferred guide in the field. The art won't inspire rapture, nor will one learn much about any one species, but the guide often includes pictures of the entire tree, rare in tree guides, and, as is usual with Golden Guides, puts maps, art and blurb conveniently on the same spread. ($11.95 paper)

Eastern Trees* (Peterson Field Guides). Probably more inclusive than the above, encompassing 455 species found in eastern North America, and containing far more information, this is probably the better guide to trees. I still used the Golden Guide in the field, if for no other reason than that the excellent art in the Peterson Guide has been removed to color plates set apart in the book's middle. (*Western Trees* also available.) ($14.95 paper)

Michigan Trees: A Guide to the Trees of Michigan and the Great Lakes Region,* by Burton V. Barnes and Warren H. Wagner, Jr. Another example of an excellent local guide, this book gives each tree species its own two-page spread, ink drawings on the right, extensive information, including endangered status, on the left. I liked this book very much, even if it did insist on using centimeters. If the book weren't so large, I'd have used it as a field guide. ($11.95 paper)

Master Tree Finder,* by the Nature Study Guild. Here's a nifty game to play with North American trees: Follow the signs from page to page until your sample leaf is identified. Sometimes you need more than this, though, to tell one species from another, but the book is a good introduction to all those green leaves. ($3.95 paper)

The Tree Identification Book,* and its companion volume, *The Shrub Identification Book,* both by George W. D. Symonds. These were invaluable identification guides for me. Although oversized and heavy, they pictured excellent, often full-sized photographs of nearly every part of the tree or shrub: leaf, bark, seed, berry, twig. These books made trees and shrubs seem much friendlier for me. ($17.95 paper each)

Guides to Other Subjects

I didn't work as extensively with other subjects, so I didn't buy as extensive a collection of guides to them. The same principles as apply to bird, tree and wildflower guides, however, generally apply to choosing a guide to butterflies, moths, mammals, mushrooms, reptiles, ferns or any other subject. I found that I generally preferred the Peterson guides, especially for mushrooms, as they provided me with excellent color plates—illustrations, not photographs—as well as a thorough description of every species mentioned.

Selected Bibliography

Babcock, Harold L. *Turtles of the Northeastern United States*. New York: Dover, 1978. Originally published in Boston Society of Natural History *Memoirs* 8, no. 3 (April 1919): 325–431.

Barnes, Burton V., and Warren H. Wagner, Jr. *Michigan Trees: A Guide to the Trees of Michigan and the Great Lakes Region*. Ann Arbor: University of Michigan Press, 1981.

Benys, Janine M. *The Field Guide to Wildlife Habitats of the Eastern United States*. New York: Fireside (Simon & Schuster), 1989.

———. *Northwoods Wildlife: A Watcher's Guide to Habitats*. Minocqua, Wis.: NorthWord Press, 1989.

Blanchan, Neltje. *Nature's Garden*. New York: Grosset & Dunlap, 1900.

Breitenback, Gary L. "Odds Against the Snapper." *Michigan Natural Resources* (September/October 1986).

Brockman, C. Frank. *Trees of North America*. New York: Golden Press, 1968.[3]

Bull, John, and John Farrand, Jr. *The Audubon Society Field Guide to North American Birds: Eastern Region*. New York: Alfred A. Knopf, 1977.[2]

Burt, William Henry. *A Field Guide to the Mammals: North America North of Mexico*. 3rd ed. Boston: Houghton Mifflin, 1976.[1]

Chambers, Kenneth A. *A Country-lover's Guide to Wildlife: Mammals, Amphibians, and Reptiles of the Northeastern United States*. New York: A Plume Book, New American Library, 1980.

Clark, William S. *A Field Guide to Hawks: North America*. Boston: Houghton Mifflin, 1987.[1]

Cobb, Boughton. *A Field Guide to the Ferns*. Boston: Houghton Mifflin, 1963.[1]

Collins, Henry Hill, Jr., and Ned R. Boyajian. *Familiar Garden Birds of America*. New York: Harper & Row, 1965.

[1] A Peterson Field Guide
[2] An Audubon Field Guide
[3] A Golden Field Guide
[4] A Golden Guide

Conant, Roger. *A Field Guide to Reptiles and Amphibians: Eastern and Central North America.* 2nd ed. Boston: Houghton Mifflin, 1975.[1]

Covell, Charles V., Jr. *A Field Guide to the Moths of Eastern North America.* Boston: Houghton Mifflin, 1984.[1]

Cumley, R. W., comp. *Fisherman's Guide to Michigan Fishes.* Houston: Professional Publication Producers, 1969.

DeGraaf, Richard M. "New Life from Dead Trees." *International Wildlife* (September/October 1978).

Ehrlich, Paul R., et. al. *The Birder's Handbook: A Field Guide to the Natural History of North American Birds.* New York: Simon & Schuster, 1988.

Elliott, Margaret Drake. *A Number of Things: The Year with Nature.* Muskegon, Mich.: Dana Printing Corporation, 1984.

Emberton, Jane. *Pods: Wildflowers and Weeds in Their Final Beauty.* New York: Charles Scribner's Sons, 1979.

Ewert, David. "Sparrows." *Michigan Natural Resources* (July/August 1984).

Ferrand, John, Jr., ed. *The Audubon Society Master Guide to Birding,* 3 vols. New York: Alfred A. Knopf, 1983.

Field Guide to the Birds of North America. 2nd ed. Washington, D.C.: National Geographic Society, 1987.

Forbush, Edward Howe; revised and abridged by John Bichard May. *A Natural History of American Birds of Eastern and Central North America.* Boston: Houghton Mifflin, 1955.

Glick, Phyllis G. *The Mushroom Trail Guide.* New York: Henry Holt, 1979.

Hallowell, Anne C., and Barbara G. Hallowell. *Fern Finder: A Guide to Native Ferns of Northeastern and Central North America.* Berkeley: Nature Study Guild, 1981.

Harrison, George H. *The Backyard Bird Watcher.* New York: Simon & Schuster, 1979.

Harrison, Kit, and George Harrison. *America's Favorite Backyard Wildlife.* New York: Simon & Schuster, 1985.

Hohn, Matthew H. *Guide to the Flora of Beaver Island: The Bogs.* Beaver Island, Mich.: Central Michigan University Biological Station, 1977.

————. *Guide to the Flora of Beaver Island, Michigan: Lake Michigan Beaches and Sand Dunes.* Beaver Island, Mich.: Central Michigan University Biological Station, 1981.

Holman, J. Alan, et. al. *Michigan Snakes: A Field Guide and Pocket Reference.* Lansing, Mich.: Michigan State University Cooperative Extension Service, 1989.

Hubbs, Carl L., and Karl F. Lagler. *Fishes of the Great Lakes Region.* Ann Arbor: The University of Michigan Press, 1958.

Lanner, Ronald M. *Autumn Leaves: A Guide to Fall Colors of the Northwoods.* Minocqua, Wis.: NorthWord Press, 1990.

Laycock, George. "Woodpecker with Pizzazz." *Audubon Magazine* (July 1985).

Levi, Herbert W., and Lorna R. Levi. *Spiders and Their Kin.* New York: Golden Press, 1990.[4]

Little, Elbert L. *The Audubon Society Field Guide to North American Trees: Eastern Region.* New York: Alfred A. Knopf, 1980.[2]

Lund, Harry C. *Michigan Wildflowers in Color.* Traverse City, Mich.: Village Press, 1988.

Lyons, Janet and Sandra Jordan. *Walking the Wetlands: A Hiker's Guide to Common Plants and Animals of Marshes, Bogs and Swamps.* New York: John Wiley & Sons, 1989.

Martin, Laura C. *Wildflower Folklore.* Chester, Conn.: Globe Pequot Press, 1984.

Maslowski, Steve. "Big Red Booms in Cincinnati." *National Wildlife* (December/ January 1983).

Mays, Verna. "Voice of the Wilderness." *National Wildlife* (December/January 1976).

McKnight, Kent H., and Vera B. McKnight. *A Field Guide to Mushrooms in North America.* Boston: Houghton Mifflin, 1987.[1]

Milne, Lorus, and Margery. *The Audubon Society Field Guide to North American Insects and Spiders.* New York: Alfred A. Knopf, 1980.[2]

Mitchell, Robert T., and Herbert S. Zim. *Butterflies and Moths: A Golden Nature Guide.* Racine, Wisc.: Golden Press, 1964.[2]

Moore, Michael D., and William B. Botti, comp. *Michigan's Famous and Historic Trees.* Midland, Mich.: Michigan Forest Association, 1977.

Nature Study Guild. *Master Tree Finder.* New York: Warner Books, 1986.

Newcomb, Lawrence. *Newcomb's Wildflower Guide.* Boston: Little, Brown, 1977.

Niering, William A., and Nancy C. Olmstead. *The Audubon Society Field Guide to North American Wildflowers: Eastern Region.* New York: Alfred A. Knopf, 1979.[2]

North American Wildlife. Pleasantville, N.Y.: Reader's Digest Association, 1982.

Pardo, Richard. "National Register of Big Trees." *American Forests* 84:18–45.

Peterson, Lee. *A Field Guide to Edible Wild Plants of Eastern and Central North America.* Boston: Houghton Mifflin, 1978.[1]

Peterson, Roger Tory. *Eastern Birds.* 4th ed. Boston: Houghton Mifflin, 1980.[1]

———. *How to Know the Birds.* New York: Signet, 1957.

Peterson, Roger Tory, and Margaret McKenny. *A Field Guide to Wildflowers of Northeastern and Northcentral North America.* Boston: Houghton Mifflin: 1968.[1]

Petrides, George A. *A Field Guide to Eastern Trees: Eastern United States and Canada.* Boston: Houghton Mifflin, 1988.[1]

Pohl, Richard W. *How to Know the Grasses.* Dubuque, Iowa: Wm. C. Brown Company, 1968.

Potemra, Thomas A. "Aurora Borealis: The Greatest Light Show on Earth." *Smithsonian Magazine* (February 1977).

Pyle, Robert Michael. *The Audubon Society Field Guide to North American Butterflies.* New York: Alfred A. Knopf, 1981.[2]

Robbins, Chandler S. et. al. *Birds of North America.* New York: Golden Press, 1983.[3]

Rupp, Rebecca. *Red Oaks & Black Birches: The Science and Lore of Trees.* Pownal, Vt.: "A Garden Way Publishing Book," Storey Communications, 1990.

Sitwell, Nigel. "A Rabbit Reader." *International Wildlife* (May/June 1983).

Smith, Alexander H., and Nancy Smith Weber. *The Mushroom Hunter's Field Guide.* Ann Arbor: University of Michigan Press, 1980.

Smith, Norman F. *Michigan Trees Worth Knowing.* 5th ed. Lansing, Mich.: TwoPeninsula Press, Michigan Department of Natural Resources, 1978.

Stebbins, Robert C. *A Field Guide to Western Reptiles and Amphibians.* 2nd ed. Boston: Houghton Mifflin, 1985.[1]

Stokes, Donald W. *A Guide to Bird Behavior.* Vol. 1. Boston: Little, Brown, 1979.

———. *A Guide to Bird Behavior.* Vol. 2. Boston: Little, Brown, 1983.

———. *A Guide to Bird Behavior.* Vol. 3. Boston: Little, Brown, 1989.

———. *A Guide to Observing Insect Lives.* Boston: Little, Brown, 1983.

———. *The Natural History of Wild Shrubs and Vines: Eastern and Central North America.* New York: Harper & Row, 1981.

————. *Nature in Winter*. Boston: Little, Brown, 1976.

Stokes, Donald W., and Lillian Q. Stokes. *A Guide to Animal Tracking and Behavior*. Boston: Little, Brown, 1986.

Swain, Ralph B. *The Insect Guide: Orders and Major Families of North American Insects*. Garden City: Doubleday, 1948.

Symonds, George W. D. *The Shrub Identification Book*. New York: Quill, 1963.

————. *The Tree Identification Book*. New York: Quill, 1958.

Terres, John K. "Masters of the Pecking Order." *National Wildlife* (December/January 1983).

United States Department of Agriculture, Forest Service. *Range Plant Handbook*. New York: Dover, 1988.

Voss, Edward G. *Michigan Flora*. Vol. 1 and 2. Bloomfield Hills, Mich.: Cranbrook Institute of Science and University of Michigan Herbarium, 1972.

Weed, Clarence M. *Butterflies Worth Knowing*. New York: Doubleday, Page & Company, 1925.

White, Richard E. *A Field Guide to the Beetles of North America*. Boston: Houghton Mifflin, 1983.[1]

A Visual Index

The show is over, but don't leave yet: It's time to run the credits. Listed below are the two hundred and twenty major participants in *Naming Nature*, grouped roughly by size, color, and/or type. Subjects within each group are arranged, very approximately, by size, with the smallest coming first. Each subject is illustrated and described in a dated entry. (The many other natural species or phenomena described in this book can be found by name in the index.)

Index

Page numbers in italic indicate main entries